A History of Camden Chinese Market Gardeners 1899-1993

Published by Camden Historical Society Inc.

40 John Street, Camden, NSW 2570

(SAN: 908 3002)

www.camdenhistory.com.au

© 2024 by the Camden Historical Society Inc.

All rights reserved. No part of this publication may be used or reproduced in any manner without written permission except for brief quotations in critical articles and reviews. Contact the Camden Historical Society PO Box 566 Camden NSW 2570 Australia for more information. (secretary@camdenhistory.org.au)

All photographs in the publication are part of the Camden Historical Society's collection or are used with the permission acknowledged in the text.

ISBN: 978-0-6485894-2-6

Front cover: Biu Wong with George Fung at Hop Chong Garden, about 1986

Back cover: 'Chinaman's Garden, Camden NSW', sketch by Douglas Annand, 1944

Contents

Contents iii
Illustrations v
Foreword ix
Preface xi
Introduction 13
1 • Camden Chinese Market Gardeners 1899-1993 17
2 • Farming Practices by the Chinese Community of Camden 21
3 • Market Gardeners' Farming Artefacts in Camden Museum 33
4 • Interaction with Camden Community 39
5 • Memories of the Chinese Market Gardeners 43
6 • George Lee 47
7 • Chun Yuen (1870-1921) 51
8 • Ah Chong 55
9 • Lowe Kum Ming and Ah On 59
10 • Wong Yong and Trips Home 65
11 • Willie Chung 67
12 • Hop Chong 75
13 • Biu Wong 79
14 • Artefacts in the Camden Museum Owned by the Chinese Market Gardeners 83
15 • Looking Back on the Camden Chinese Market Gardeners 89
Conclusion 95
Appendix 101
Bibliography 103
Acknowledgements 105
Author Biographies 107
Index 109

Illustrations

Figure 1 - Guangdong Province.	17
Figure 2 - Yiu Ming Temple, a Chinese temple, Alexandria, Sydney NSW (John Wrigley, 2015).	19
Figure 3 - Aerial photo of Camden looking west in the late 1930s (Camden Museum). Note: Chinese market gardens in the foreground.	20
Figure 4 - Locations of the Chinese market gardens (Nixon).	22
Figure 5 - Bond Brothers' truck taking Chinese produce to Sydney market, 1964 (Mrs Lyn Luscombe, nee Bond).	27
Figure 6 - Willie Chung, Immigration papers, 1946 (National Archives of Australia ST84/1, C1946/5927, Page 97/97).	29
Figure 7 - Camden view looking along John Street from the steeple of St. John's. The house of the Davies' market gardeners (Tong Hing garden) is visible in the far distance near the river. The site of the vegetable crops is currently part of the Miss Llewella Davies Walkway.	30
Figure 8 - Dianne Holmquist nee Wong (dark blue jumper) and Holmquist family visiting Camden Museum, 10 June 2019 (photo by Roslyn Tildsley).	32
Figure 9 - Watering cans at Camden Museum. Items 2012.26 and 27 (John Wrigley, 2014).	33
Figure 10 - Two of the steel rakes which were used by the gardeners. Item 2012.25.	35
Figure 11 - Miss Davies' Tong Hing Rent Receipt Book for 1962-3. Item 2022.16 (Anne McIntosh).	36
Figure 12 - Davies' Market Garden lease, 1946. p. 2 (Camden Museum).	37
Figure 13 - 'Chinaman's Garden, Camden NSW', sketch by Douglas Annand, from the book Drawings and Paintings 1944 (Item 2022.18.).	38
Figure 14 - The Chinese were rescued and accommodated in the Town Hall (*Daily Telegraph* 23 May 1943).	41
Figure 15 – A rare photo of Camden Chinese gardeners washed out by a flood in 1925 (Camden Museum).	42
Figure 16 - The Chinese garden sheds were flooded in 1943 (Mrs Lyn Luscombe. nee Bond).	44
Figure 17 - Chinese Gardens, Grove Road, Camden, near the Macquarie Grove treatment works. Miss Davies' paddocks in the background, Jack Dunk driving the tractor, Stan Rofe watching (Joy Riley).	45
Figure 18 - Jack Dunk on the tractor, preparing the soil for Chinese market gardeners, Stan Rofe standing (Joy Riley).	45
Figure 19 - Ho Jo Ling gave this beaded bag to Joy Riley when she was a girl.	46
Figure 20 - George Lee, 1910 extracted from his Certificate Exempting from Dictation Test (National Archives of Australia: ST84/1, 1910/44/1-10. Page 9).	49

Figure 21 - George Lee's Certificate Exempting from Dictation Test, 1910 National Archives of Australia: ST84/1, 1910/44/1-10. Page 9). — 50

Figure 22 - Chun Yuen's Certificate Exempting from Dictation Test, 1909 (National Archives of Australia J2483, 17/30. Page 1). — 52

Figure 23 - Chun Yuen's left-hand print on his Certificate Exempting from Dictation Test, 1909. (National Archives of Australia: J2483, 17/30. Page 2). — 53

Figure 24 - Death Certificate for Chun Yuen. — 54

Figure 25 - Camden-Campbelltown train on Cowpastures Bridge, 1910 (Camden Museum). — 56

Figure 26 - Camden-Campbelltown train, Pansy, about 1917 at Camden Railway Station (Harold Perkins). — 57

Figure 27 - Camden-Campbelltown locomotive Pansy and the loading siding near the former Milk Depot. Note how dangerously close the train lines are to pedestrians (Camden Museum). — 58

Figure 28 - Ah On, Immigration papers, 1922. (National Archives of Australia: Certificate Exempting from Dictation Test ST84/1, 1922/333/91-100. Page 7). — 59

Figure 29 - Innocent victim Kum Ming, 1932. (National Archives of Australia, NSW Office Chester Hill SP42/1, C1932/2547. — 61

Figure 30 - Rookwood, Lowe Kum Ming headstone lower right corner (John Wrigley, 2016). — 63

Figure 31 - Archie Tippetts' truck in Elizabeth Street, laden with cauliflowers, grown by local Chinese, 1932. One truck has the sign "AC Tippetts Camden – Sydney" (Camden Museum). — 65

Figure 32 - Chung Wong Ying also known as Wong Yong (Daphne Lowe Kelley). — 66

Figure 33 - Camden District Hospital Life Members Wong Yong (Tong) and Willie Chung (Chong) (John Wrigley, 2017). — 67

Figure 34 - Willie Chung, Immigration papers, 1912 (National Archives of Australia SP42/1 C1946/5927, page 85/97). — 68

Figure 35 - Doctor's letter confirming Willie and Charlie Chung were 'half caste' (National Archives of Australia: SP42/1 / C1946/5927, page 50/97). — 69

Figure 36 - Willie Chung's Certificate Exempting from Dictation Test, 1922 (National Archives of Australia: Willie Chung ST84/1, 1922/333/91-100). — 71

Figure 37 - Mayor Kelloway's letter of support (NAA: SP42/1, C1946/5927 page 5/97). — 72

Figure 38 - Grave of Willie Chung at Rookwood (John Wrigley, 2016). — 73

Figure 39 - Willie Chung, 1937 (National Archives of Australia). — 74

Figure 40 - Yung Sum Wong and Siu Wong (the parents of Biu Wong) with their granddaughters Anne and Dianne and Sim and Biu Wong at Hop Chong Garden, 1969 (Photo supplied by the Wong family to Camden Museum). — 77

Figure 41 - Sim Wong and baby Anne at Hop Chong Garden in 1967 (Photo supplied by the Wong family to Camden Museum). — 77

Figure 42 - Sim Wong in Hop Chong Garden Shop, Camden, 1980s (Wong family). — 78

Figure 43 - Hop Chong Garden Shop, 3 Argyle Street, Camden, 1990s (Wong family). 78

Figure 44 - Biu Wong, with George Fung, Sim Wong's brother-in-law, who was visiting from Hong Kong, 1986 (Photo supplied by the Wong family to Camden Museum). 79

Figure 45 - Wedding of Biu and Sim Wong. St John's Church, Camden, 1965 (Wong family). 80

Figure 46 - Sim and Biu Wong inside the Hop Chong Garden Shop, 1980s (Wong family). 81

Figure 47 - Biu Wong on his Massey Ferguson tractor, Hop Chong Garden, 1980s (Wong family). 82

Figure 48 - Sim Wong and Andrew on the same Massey Ferguson tractor, 1980s (Wong family). 82

Figure 49 - Wong family visiting the Nixon Room at the Camden Museum, 2022 (Andrew Lui). 83

Figure 50 - Abacus. Item 1980.1 (Anne McIntosh). 84

Figure 51 - Chinese hat. Item 2012.24 (Anne McIntosh). 84

Figure 52 - Framed silk embroidery. Item 1980.283 (Anne McIntosh). 85

Figure 53 - Dragon bowl. Item 1995.449 (Anne McIntosh). 86

Figure 54 - Rice Wine bottles. Items 2015.12 and 1980.186 (Anne McIntosh). 87

Figure 55 - Wong family in the Chinese bay of the Camden Museum, 2022 (Andrew Liu). 88

Figure 56 - Wong family in the Research Room at the Camden Museum, 2022 (Andrew Lui). 88

Figure 57 - Ho Jo Ling, Foon Kee Pan and Ben Pan as a boy, c. 1947. 90

Figure 58 - Sketch of Tong Hing market garden on Davies' farm, hand-drawn by Ben Pan, 2023. 91

Figure 59 - Wong family with John and Julie Wrigley at Camden Museum, 2022 (Andrew Lui). 93

Figure 60 - Biu Wong, Sim Wong and John Wrigley at the Camden Museum, 2022 (Andrew Lui). 93

Foreword

Camden has had a long, proud, rich, and diverse history. From the Aboriginal nations who lived and traded in its plains through to the colonial and immigrant residents who moved into the area for the rich soils and vast grasslands, Camden has held people's imaginations through generations.

This history could easily be lost from the memories of the current inhabitants if it was not for the enduring work of many people such as Ian Willis, Julie Wrigley and the Camden Historical Society. The importance of sharing these histories cannot be underestimated for their enduring influence on the nation's narrative. Whether it is indigenous, colonial, agricultural or migrant stories, all have a part to play in the fabric of who we are.

Camden has a strong storytelling culture. The use of stories to help preserve and humanise history, making it more accessible and interesting, allows this history to be remembered and understood. This in turn means that current and future generations will have the opportunity to explore and experience this history, allowing its lessons to endure.

In this great book, Ian Willis and Julie Wrigley have done an excellent job of bringing forth the stories of a small, but very important part, of our region's history, the story of Chinese market gardeners in Camden. As Julie notes in the story of local Chun Yuen, although the Chinese market gardeners had a long and important link with Camden, their story was only told sporadically and through the lens of other people's perceptions.

This work seeks to remove the lens of other people's viewpoint and allows the story of Chinese market gardeners in Camden to be told from the point of view of the gardeners themselves.

The result of this work is a comprehensive encapsulation of a moment in time, from the viewpoint of those experiencing it. A collection of important stories, which allow us a window into one fabric in the quilt of our region. We can reflect how these residents have influenced the generational history of our region, and Australia as a whole, allowing us to see how we have journeyed to become the Camden we are today.

Sally Anne Quinnell

Member of the NSW Legislative Assembly, Member for Camden

Preface

Julie Wrigley

Camden has always had a strong sense of community, and the Camden Historical Society has been collecting objects to remember the Chinese market gardeners since the Camden Museum opened in 1970. In 2016, local China-born resident Wen Denaro came into the Camden Museum and asked questions about the history of Chinese people who had lived in Camden. I started researching and read the article by Richard Nixon written in 1976. I found that the combination of the local newspapers in Trove and records in the National Archives of Australia revealed many stories about the Chinese men who had lived and worked in Camden. I published several articles in the local newspaper, *The District Reporter*. In 2017, Wen Denaro obtained a grant to produce a DVD, *100 Yards of Silk*, about the difficulties endured by the Chinese market gardeners in Camden. At the same time, I produced a PowerPoint presentation, *Chinese Market Gardeners in Camden from 1899 to 1993*, for the Camden Historical Society. The PowerPoint presentation and other research included contributions by living Australian Chinese descendants, telling the story of the Camden Chinese by the Wong family. Genealogy and family history play an essential role in this publication's compilation, which draws together details of other Camden Chinese, assisted by sources like births, deaths, marriages, and immigration records.

In 2022, Sydney University historian Sophie Loy-Wilson visited the Camden Museum. She watched the DVD and PowerPoint presentation and looked through the archive box on the Chinese market gardeners. She urged the Camden Historical Society to gather the articles into a book. Sophie Loy-Wilson consulted with Ian Willis, a local historian and academic who has written three books on Camden's history. Ian drew up an outline of how the material could be brought together and supervised the book's preparation, advising on the whole process.

The book's first part focuses on the farming practices of the Camden Chinese market gardeners and explains how they contributed to Australian agriculture. This includes the article written by Richard Nixon in 1976. He explains how the Chinese cooperatives operated and outlines the cultivation and irrigation methods the Chinese market gardeners used to produce vegetables for the local and Sydney markets. In this section, John Wrigley has written a chapter on Museum items to do with the agriculture of the Chinese market gardeners, telling stories about the objects that reveal the farming practices of the market gardeners. Another chapter is about the interaction between the Chinese farmers and the Camden community, and the respect the Chinese market gardeners gained. It includes the memories of the Chinese market gardeners from residents Miriam Dunn and Joy Riley.

The book's second part considers the lives of the Camden Chinese market gardeners and explains how they contributed to settlement and community. This section deals with the lives of individual Camden Chinese market gardeners, their high and low points, unfortunate accidental deaths and the tragedy of a murder and suicide. The chapters show how conditions changed from George Lee in the 1910s to Chun Yuen and Ah Chong in the 1920s, Lowe Kum Ming and Ah On in the 1930s, Wong Yong and Willie Chung in the 1940s, and the Hop Chong Garden story from the 1960s. Immigration regulations changed, and the experiences of the Chinese market gardeners changed.

Preface

John Wrigley has written an article about items in the Camden Museum owned and used by the Camden Chinese market gardeners. The objects reveal aspects of the way of life of the gardeners. There is a chapter in this section on the last of the Camden Chinese market gardeners, Biu Wong, and a consideration of the contribution of the market gardeners to Camden's history and culture.

In Australia, there has always been confusion about Chinese surnames. Australian officials assumed the name order was a personal name followed by a family name, but this was not Chinese practice. In Chinese, the family name comes first, then the personal name. Also, one person might be referred to by several different names at different times or to show different levels of respect. The authorities should have been more careful with names, which were often misheard or ignored, but they were strict in applying the discrimination of the legislation. For the Chinese gardeners in Camden, I have followed the order of names used by the authorities, including the National Archives of Australia and newspapers in Trove.

In 2023, the Camden Historical Society decided to publish this book with the help of Fletcher Joss as project manager. Gathering the material into a book would only have happened with the interest and enthusiasm of Sophie Loy-Wilson, who kindly agreed to write the introduction. Thank you to Sally Quinnell, Member of the NSW Legislative Assembly, Member for Camden, for writing the foreword.

This book will help readers understand the hardships the Camden Chinese market gardeners experienced in the past and help build good relations between China and Australia in the future.

Julie Wrigley

Introduction

Sophie Loy-Wilson

University of Sydney

This book tells the story of Camden's Chinese market gardeners. Julie Wrigley has collated years of research to bring us this tale. It is a story which takes the reader from the outskirts of Sydney to rural China, to Hong Kong and back again, from the turn of the century, to the 1990s. It is a book which shows, indisputably, that Chinese farmers helped make this country; that their work was at the heart of Australian urbanization.

I don't know what it means to farm, but recently I tried to imagine it: the early rising, the sweat, the thirst, the hunger, the stinking manure, the aching muscle and bones, the endless, endless watering, the pride in growth and harvest, the satisfaction brought by success, when it came, if it came. But then also the intense fragility of it all – the psychological awareness that one might lose everything and fast: through flood, through storm, through infestation, through the God-like power of the very natural forces farming seeks to harness and control. The biblical tragedies that come with putting your faith in the land, in the soil.

What must this have been like? For the Chinese Australian farmers whose lives I study, all this risk was shot through with other fears. For the land was not there to nourish the gardeners alone, but also their impoverished family back home; wives, children, parents, sometimes whole villages – hopeful, vulnerable dependents across the seas – their lives hung in the balance as well, as the rain fell.[1] Or didn't. Or too heavily. What must this have been like when not only was the soil hostile but the swelling city as well, populated by mostly white settlers increasingly bent on rendering life for Chinese farmers in Australia, uncomfortable, unviable, and dangerous? The University of Sydney Rare Books Library recently acquired the draft of a proposed bill put to the NSW Parliament in 1881 to better protect Chinese from 'ill-treatment and assault' and detailed the extremities of these attacks, especially on farmers, vegetable hawkers.[2] Anti-Chinese violence stalked these farmers and beatings from youth gangs of organized anti-Chinese leagues were always just on the horizon; they lived with this fear. The bill didn't pass.

Suburban Sydney, where I live, is pockmarked by old farms – we call them market gardens – but they are farms, vegetable farms, and they are old, they date back to the 1860s, and they bear the marks of many peoples, many cultures.[3] Chinese fruit trees were placed at the back of mangrove swamps, famously hiding places for Indigenous peoples fleeing white violence and harassment. Greek and Italian grape varieties still push through the soil next to lotus flowers and old bamboo stalks and banana trees. The hands that planted them would have been rough-worked, as they say, to the bone, and tanned by the Australian sun, which beat down so relentlessly that some gardeners

---

[1] Michael Williams, Returning Home with Glory: Chinese Villages around the Pacific, 1849 to 1949 (Hong Kong: Hong Kong University Press, 2018), p 4.

[2] 'A Bill for the better Protection of Chinese from insult ill-treatment and assault. [Short title] Chinese Protection Act of 1881,' Rare Books Library, University of Sydney.

[3] David M George, 'The Celestial Question: Chinese communities in the Bayside Area in the late 19th and early 20th centuries,' Winner of the 2019 Ron Rathbone Local History Prize, Bayside Local History Library Collection.

Introduction

had to cover all their produce with straw, one of many techniques learnt in an intimate conversation with a land they barely knew, that they were trying to know, to survive.[4] Perhaps Eora people watched on, from within the mangroves. These gardens are now boxed in by highways, houses, even an airport – but their straight lines survive, and if you walk along these lines, produce still rises high, and the insects still buzz over the brackish water long used to irrigate, to nourish, to bring food to a hungry, avaricious city in the world's most urban country; the settler colony we know as Australia.

Life in Australia was multicultural long before the term was in common use.[5] In Camden British, Irish, Scottish settlers, often sunburnt, toiled long hours as farm labourers, working alongside itinerant Indigenous corn pickers, German wine growers lured by the gold rush and a community of Chinese market gardeners who settled in the area in the 1890s. Their population peaked between the wars, the last Chinese market gardeners selling up in 1996, telling their children to choose an easier life.[6] Early European arrivals knew hardship, they were mostly convicts. Irish migrants had survived the Great Famine, some British were rebels from the 1830s riots. At first Sydney seemed far, very far – Camden was exile layered upon exile – but then roads were cut into the land, coaches ran, and finally trains, meaning Sydney as market, as centre of order, became a vein of capital, and some families did well, although farming by the banks of a volatile river was forever a gamble. Crops were often disseminated, children drowned, farms abandoned. Such was a life shared with time and the river.

Camden is now a municipality sixty-five kilometers southwest of Sydney, about a forty-minute drive, with a population of 2,300 residents. It also has a thriving historical society, responsible for the preservation of not only Camden's European histories, but Indigenous and Chinese as well. This book is a testament to the broad Church that is that local society, and the many Camden treasures kept safe within its walls by Julie and John Wrigley and others: Chinese watering cans preserved by the Italian families who inherited Chinese farms in the post war period, steel rakes, ginger jars, rent receipt books, lease agreements bearing Chinese signatures, silk embroidery and purses, an abacus. Photographs from flood times capture market gardeners in situ, at points of rescue and relief. The drawings from Australian war artist, Douglas Annand, stationed at Camden in 1941-44 with the Australian air force, capture Chinese market gardeners at work. Perhaps most precious of all are the memories; oral histories collected since the 1970s richly document this multi-ethnic community, its ebbs and flows, its accommodations, and curiosities.

Floods were in everyone's memories, an 'apprenticeship in misery' in Alan Atkinson's words. A story left by Boodbury, one of the Camden Aborigines told of a flood when he was a little boy, probably the great flood of 1806.[7] The river was unconquerable, and remains so to this day, drawing life to its banks, feedings crops – long rows of meticulously tended Chinese gardens – but just as forcefully destroying it all overnight, whole harvests ruined. One needed a gambling spirit to work irrigation on such a river, and Chinese farmers, by necessity but also predilection, were keen gamblers. The

---

[4] Julie Wrigley, *A History of Camden Chinese Market Gardeners* (Camden: Camden Historical Society, 2024) p 15.
[5] Sophie Loy-Wilson, 'Introduction,' in Mavis Gock Yen, *South Flows the Pearl: Chinese Australians in White Australia* (Sydney: University of Sydney press, 2022), pp 9-10.
[6] Wrigley, *A History of Camden Chinese Market Gardeners*, p 8.
[7] Alan Atkinson, *Camden: Farm and Village Life in early New South Wales* (Oxford: Oxford University Press, 2008) p 94.

river also made men heroes – one Camden local learnt how to make bark canoes the Indigenous way, rescuing many townsfolk in the floods of 1857.[8]

For Indigenous peoples, the area Europeans called Camden was likely Baragil or Baragal; the Gundungarra people, perhaps with a touch of irony, called it 'Benkennie' or dry land.[9] We don't know how Chinese settlers referred to the land they worked so closely, its clay soil dying all hands a bloody brown and turning to glue in the rain. Australian history has generally relied on English-language voices and memories, and so the dense webs of cultures and tongues that marked life in Camden has rarely been documented; this book is an exception, and a valuable one.

These were the White Australia years, that strange cruel biological experiment designed to keep the nation white, and non-Anglo Australians were actively excluded from the offerings of our newly formed Australian democracy – from the right to vote, buy land, bring in family, access welfare, travel freely.[10] And yet many stayed here, worked and died here, leaving their stories behind on the land and in the minds of those who knew them; memories reproduced in the oral history transcripts in this book. In Camden, these imprints are deep – deep as the irrigation trenches running through the old Chinese market gardens, deep as the wells dug to absorb flood water, deep as the banks of the Nepean River, which claimed men's lives, some Chinese, and in which Chinese migrants ran to hide during raids by immigration officials in the 1930s. The river gave to the town and took from the town in equal measure. It flooded frequently and violently over years, and these floods attracted photographers who captured occasions of recue; Chinese farmers ferried to safety by locals, sometimes at night, their torches flashing across the water as they signaled their distress from rooftops. They were rescued and given refuge in the homes of their landlords at the local Agricultural Hall.[11] The drama of the display, set against the peace of the surrounding fields, suggested the extremes of human existence.

Chinese market gardeners at Camden rose early; sometimes 5 a.m., and they worked till late, until their cook raised a red flag over the gardens, and the men walked, sometimes by lamplight, in single file on narrow pathways, into their lodgings for dinner. Their vegetables were vulnerable, and the gardens were meticulously organized so the men could avoid walking through their produce, damaging young bean sprouts, for example, or precious tomato produce, carefully covered in straw to protect the tomatoes from the Australian sun. They could not protect themselves from pilferers – especially during the Great Depression – and they used guns to frighten off birds, often cockatoos whose loud cries so startled the early white invaders. An early botanist remarked on how much the noise hurt him.[12] As the men walked into dinner, a series of calculations likely ran through their minds: the price of manure, the price of crop transport to market, the likely profit they might make from their produce, the likelihood of bad weather, the exchange rate with China, the value of their money once it finally landed in the hands of their family back home. If we see this world through their eyes, then the village in China is always present – homes, wives, parents, children – all come into view.[13] As Nixon writes, "During the Sino-Japanese War, they were intensely loyal to the land of their birth, sending home money for the war effort. One remembers one Chinaman leaving for

---

[8] Ibid, p 209.
[9] Ibid, pp 7-8.
[10] Loy-Wilson, 'Introduction' p 3.
[11] Wrigley, Camden Chinese Market Gardeners, p 29.
[12] Atkinson, Camden, p.6.
[13] Williams, Returning Home with Glory, pp 8-11.

Introduction

home, and his one aim was to take with him sufficient money to buy himself a gun to use against the Japanese." [14]

In Australia, the importance of large-scale pastoralism and agriculture for export or mixed farming meant that small scale, intensive market gardening as the sole source of income was considered of low status. This opened an economic niche that Chinese labourers filled, lured by the gold rush, but aware of the high prices for vegetables. In rapidly growing towns and cities, food, and food delivered in a timely manner is vital. During the pandemic lockdowns, how and where we ate changed, drawing ever more attention to the people who feed Sydney today. We would do well to remember these early gardeners and the sustenance they provided us. Julie Wrigley has ensured that we will.

---

[14] Nixon, Camden Chinese Market Gardeners, p 20.

1 • Camden Chinese Market Gardeners 1899-1993

Julie Wrigley

Only a few people know the story of the Chinese market gardeners who lived and worked in the riverside town of Camden in New South Wales, and they deserve to be remembered. From the late 1890s to 1993, there was a community of Chinese market gardeners in Camden. They endured physical hardship in working long hours, droughts and floods, loneliness away from wives and families, isolation in language and culture, and government discrimination through the *Immigration Restriction Act 1901* [15] and the strict regulation of the Certificates Exempting from the Dictation Test [16]. The men were sustained by support from other members of their clan or family, the satisfaction of growing and selling vegetables and making enough money to send home to families in China, and goodwill from the Camden storekeepers and carriers who supplied food and services.

At Camden Museum, there is a bay remembering the Chinese market gardeners who came to Camden in the early twentieth century. The Chinese were nearly all men with their families still in China, working in small cooperatives along the banks of the Nepean River growing vegetables for the Sydney market. Many men were cousins or relatives from the same district and spoke the same dialect. Almost all the Chinese market gardeners who came to Camden in the early twentieth century were born in the Guangdong Province, near the city of Guangzhou (formerly known as Canton) [17] and spoke Cantonese.

Figure 1 - Guangdong Province.

---

[15] The Immigration Restriction Act (White Australia Policy). A law passed by the Commonwealth of Australia that allowed immigration officials to discriminate against people according to their skin colour.

[16] After 1905 Chinese in Australia wishing to travel back to China had to apply for a Certificate Exempting from Dictation Test to be sure they could return to Australia without having to submit to a diction test which could be in any language.

[17] https://www.chinahighlights.com/guangdong/map.htm

1 • Camden Chinese Market Gardeners 1899-1993

It was previously thought that the Chinese market gardeners worked in Camden from about 1910, but new research in Trove has revealed that in 1899, George Lee, a Chinese market gardener of Elderslie, had 'the first Chinese garden in this district.' [18] His story follows later in the book.

The *Camden News* (1 Dec 1910) praised the skill of the Chinese market gardeners:

> It is very interesting to watch those enterprising and skilful agriculturalists who are working the vegetable gardens near the Camden Bridge. They seem to be thoroughly conversant with the composition of the soil and whatever deficiency there was, they spared no expense in making it good…We learn that from £400 to £500 have been expended on the gardens on the right hand side of the road going over the bridge, but the tenants will more than get back their money in a short time…The Chinamen on the flats are now sending to market from two to three tons of tomatoes weekly. They were selling the tomatoes locally as high as one shilling and sixpence a dozen.

Hard Work on Fertile Soil

One of the best descriptions of the gardening scene in these early days is from the *South Coast Times and Wollongong Argus* (8 December 1911) with the heading 'Successful Chinese Gardeners'.

> The Chinese gardens on the river flat at Camden are worthy of notice, and a strong evidence of what intense cultivation will do. One of the gardens, twenty-two acres in extent, is run by a syndicate of eight Chinese (all workers), who manage this area by themselves in the winter months, and in the summer time employ six Chinese labourers at a wage of thirty shillings a week and their keep. At the present time the whole of their leased land – for which they pay an annual rental of £132 (or £6 per acre rental) is practically covered with vegetables of every description. With this land is leased an irrigation plant, and what with water and plenty of manure, the industrious Celestials are able to raise truck-loads of vegetables for the metropolitan market. The Chinese are born gardeners and irrigationalists, and thoroughly understand the value of rotation of crops. No wonder the returns are remarkably good…It is safe to assume that these men from the Flowery Land will amass sufficient money within ten years to keep them in affluence in China for the rest of their lives.

Note that the article praises the industry and ingenuity of the workers but uses condescending words like 'the industrious Celestials' and 'the men from the Flowery Land', labelling the gardeners as outsiders.

Another article in the *Camden News* (7 November 1912) stated:

> We reported a short time back that several Chinese had started market gardens in Camden, and all had irrigation plants. On enquiry at the [railway] station last Monday, we learned that 61 tons of vegetables, mostly cabbages, were trucked. Each truck will carry about five tons of vegetables. Three times a week, Chinese and local farmers of other nationalities send enormous quantities of vegetables to

---

[18] Camden News 16 March 1939, looking back 50 years. The Camden News was a weekly newspaper published in Camden, New South Wales, from 1881 to 1982.

the Sydney market. Over 200 tons of produce were trucked at Camden last Monday, and enormous pressure was placed on the few hands who had the handling and looking after the goods. The station is clearly undermanned and the fact should at once be made known to the Commissioner for Railways.

The market gardeners worked hard, using hand-made tools in those early days, from sunrise to sunset almost every day of the week. They earned more money than they could in China, earning enough money to send money home to their wives or family and to go home every four or five years.

They were law-abiding and polite and treated with goodwill by the townsfolk. Still, there were barriers of language and culture, and only those who could speak English formed friendships – the boss or the buyer who had a relationship with the baker, the butcher, and the carriers. The men lived in dormitories in simple sheds and kept to themselves, but when they were ill, they were treated by the local doctors and Camden District Hospital. From time to time, the men joined in community activities such as fundraising and buying lottery tickets, and sometimes they brought gifts when they returned from trips to China for the children of the landowners. They used the Camden-Campbelltown train, affectionately known as Pansy, to sell vegetables to Sydney markets and make occasional trips to Sydney to visit Chinatown, Haymarket or Chinese temples in Glebe or Alexandria. [19]

Figure 2 - Yiu Ming Temple, a Chinese temple, Alexandria, Sydney NSW (John Wrigley, 2015).

In the early days, district organisations were established to assist the gardeners' fellow clansmen. Luen Fook Tong was an organisation established in Sydney's Chinatown for those from Jung Seng (Zengcheng) County, Guangdong Province, China. [20]

---

[19] Yiu Ming Temple, 16-22 Retreat Street, Alexandria, Sydney NSW Australia. Built in 1908–9, it is one of the oldest surviving Chinese temples in Australia.
[20] Unpublished notes from Daphne Lowe Kelley for Camden Historical Society, 12 July 2017.

1 • Camden Chinese Market Gardeners 1899-1993

The land was leased from owners who drew up a year-long contract for the Chinese men to sign. If they returned to China, they could sell their share in the cooperative. There were occasional conflicts over cards or gambling debts, but very few arguments. The Chinese market gardeners came for economic reasons. They were not coming as individuals to make themselves rich but to send money to their families in villages in China and to return to their villages at the end of their working life. Many arrived in debt, owing money to the agents or companies who had helped them come. The Chinese gardeners worked on their gardens, living side by side with the residents, but remained outside the mainstream Camden community.

Conditions changed for the market gardeners in Camden before and after the Second World War. The right of Australian citizenship was finally granted to Chinese people in 1958, and in several steps, the White Australia Policy ended, and government restrictions were lifted. The gender imbalance changed, and Chinese men could finally bring their families. The Greek and Italian communities started to move into farming and growing vegetables, particularly after the war.

The last Chinese market gardener in Camden was Biu Wong, who purchased the Hop Chong Company in 1968 and gardened there with his family until 1993. Biu Wong and his wife Sim were greatly respected in the community, but Biu said the life of a market gardener was a life of hardship and long hours.

The Chinese market gardeners have gone from Camden, but the historical society believes that Chinese history in Australian towns should be remembered.

Figure 3 - Aerial photo of Camden looking west in the late 1930s (Camden Museum). Note: Chinese market gardens in the foreground.

2 • Farming Practices by the Chinese Community of Camden

Richard Nixon (1976)

The Market Gardens

Camden Chinese vegetable market gardens were conducted by groups of about twelve Chinamen in six prominent locations, and in addition, there were one or two small family units. Each group's number varied with the ground under cultivation area – usually 10 to 20 acres. The groups were known locally, first by the name of the owner of the land on which they farmed and second by their Chinese 'Company' name:

1. *Nesbitt's* Chinamen [21] were located on the eastern side of the Nepean River, about 500 metres south of the present Macarthur Bridge across the Nepean. *Sing Mo* led them.

2. *Thurn's* Chinamen were also on the eastern side of the river, adjoining the above. The Macarthur Bridge passes over what was part of this garden. *Hop Chong* was an early leader.

3. *Church or Town* Chinamen were on the western bank of the river, bounded by the old Hume Highway and the 'Rectory Paddock'. This was to the south of the old highway. This group had no connection with any Church, but their garden was close to St. John's, hence the name. 'Town' referred to their closeness to Camden. *Yee Lee* was the leader.

4. *Whiteman's* Chinamen were located on the western bank of the river, bounded on the south by the old highway and the then railway line. Young Lee, sometimes known as Yong Lee, was the leader. Later San Yeck took over here, and later still *Hop Chong*. This group name was known in the district for something like 60 years. The last Chinese occupant, Biu Wong, took over in about 1970.

5. *Sheil's* Chinamen were located on the eastern bank of the river, adjacent to Camden Weir, and bounded by Narellan Creek on the eastern side. *Sun Chong Kee* was the leader for thirty years.

6. *Davies'* Chinamen were located on the western side of Macquarie Grove Road, south of Macquarie Grove Bridge, and bounded by the Nepean River on the north. *Tong Hing* was the leader for something like 50 years.

A small family unit was located on the eastern side of the old highway at Elderslie, about mid-way between Hilder Street and Wilkinson Street, in the late 1920s. Another small unit was located on the western bank of the Nepean, adjacent to Camden Weir, on what could be considered an extension to Mitchell Street.

---

[21] Nixon is using language of the 1970s referring to the Chinese as Chinamen.

2 • Farming Practices by the Chinese Community of Camden

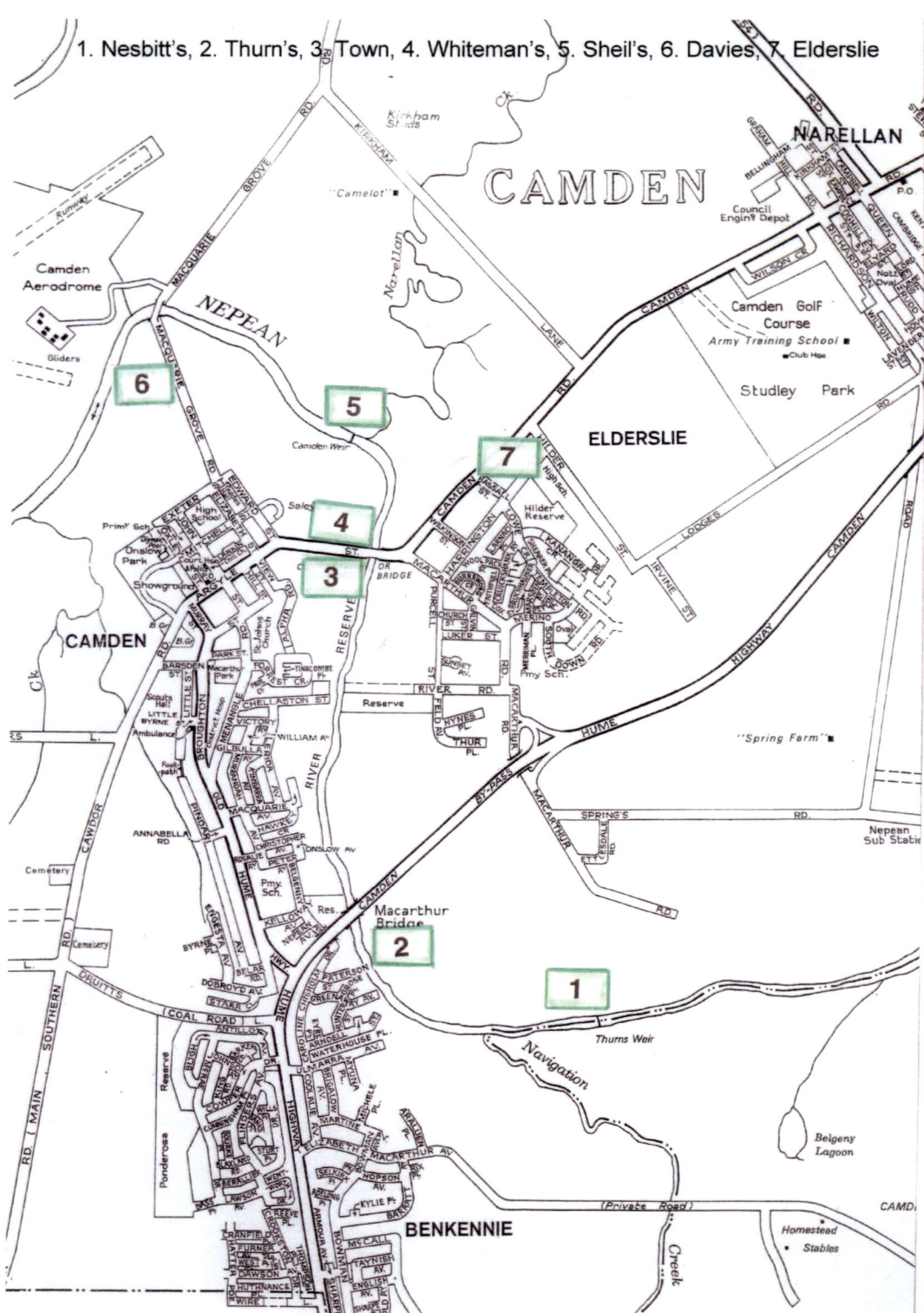

Figure 4 - Locations of the Chinese market gardens (Nixon).

Each unit operated on a cooperative system – each member contributed a share of the capital and took a share of the profit or the loss. This agreement between themselves seemed to have lasted for a year at a time, from one Chinese New Year to the next.

Each garden had a leader, or boss, who conducted the business. The locals called these men Charlie, George, or Willie. I think it would be fair to say that very few, if any, residents took the time or trouble to learn the names of any of the Chinamen. The group leader at each garden was known locally as the 'boss' of the garden. Evidence suggests he was the leader because he had the most financial share. Usually, he was a man of outstanding character. The leaders, together with the cooks, were probably the only Chinamen most of the locals ever got to know.

The next most important member of the group was the 'market man', who sometimes was the boss. The market man was the man who went to Sydney on the train, usually about 4 to 4.30 p.m. and supervised the selling of the produce at the Sydney produce market. He returned by train the following day, usually just before lunch.

The scheme of things made the 'engine man' number three on the list. He looked after the irrigation pumping plant on the riverbank. His importance declined in the 1930s with the introduction of electric pumps, and steam-driven pumping ceased. The number four man on the team was the 'cook', who became well known to local shopkeepers as he made his daily purchases. There did not appear to be any general fraternising of one garden group with another. Social 'get-togethers' as such did not exist. They only got together on Chinese New Year, when they let off fireworks, and in flood time, they were billeted in the AH&I Hall. [Also called the Agriculture Hall or Show Hall].

Housing and Diet

Their houses generally were constructed with corrugated galvanised iron because of flooding, with separate small bedrooms opening off a central hallway. There were wooden floors in the sleeping area and dirt floors in the kitchen and eating areas.

Cooking was done in a fireplace over an open fire, using Chinese cooking utensils – usually suspended on chains. Much food was cooked in the wok. When the meal was ready, the cook flew a flag on a pole attached to the house. The flag was a square of red cloth, usually red turkey twill, about three feet square. This signalled the workers to return from the garden to the house. The meal would be eaten around a wooden table using chopsticks from Chinese bowls. Pork and poultry also formed part of their diet, with other Chinese foods usually brought back from Sydney by the marketman.

Purchases of perishable foods had to be made daily. Their main food was rice, boiled. In the early days before rice growing was established in New South Wales, most of the rice eaten was imported from China. It came packed in bags of about 60 to 70 lbs, made of woven straw. The bag was made of a rectangle of woven rice straw, three feet by two feet, folded over, with the top, bottom, and side sewn together with a cord. Opened flat, the bag made a valuable and decorative floor mat, and many a local home during the 1930s depression years used these as floor mats. In addition to eating rice, they fed rice to their horses, dogs, cats, and poultry, and I never saw one of their animals in poor condition. China tea was probably their main drink, with Chinese whisky and Australian beer as their main alcoholic drink.

2 • Farming Practices by the Chinese Community of Camden

Flooding

Because of their locations on the banks of the Nepean River, the Chinamen were always victims of flooding. Time and time again, crops of vegetables would be lost to floods. Garden beds would be damaged, and floodwaters would pass through their houses. The change from steam to electrically driven irrigation pumps aggravated the pump damage problem. To the best of my knowledge, no relief funds were ever made available to them, but despite their losses, come New Year, all debts were paid. During flood periods, they would move into the AH&I Society Hall but were always keen to return to their home as soon as possible – sometimes before it was possible!

On the other hand, it is true to say that the major floods of 1949, 1954 and 1956, the devastation they caused to the gardens, contributed to their demise. From 1939 to 1945, this district suffered virtually one long continuous drought. This was when the river dried up, and it was possible to walk for miles up or down the dried up sand of the riverbed. One should remember that at most fifteen irrigation pumps were on the river at this period, from Menangle to Cobbitty! This gives some idea of the severity of the drought. To keep their gardens growing, they dug 'soak holes' in the riverbed, with drains trying to lead any water into the soak hole around the pump inlet. Heavy rain fell from May to June 1943, and indeed, there was a flash flood that closed Camden Bridge for a few hours, but for the remainder of the year, almost no rain fell. 1944 or 1945 were drought years, and with the entry of Japan into the 1939-45 war, in December 1941, many Chinese left the district virtually overnight, probably fearing they would be mistaken for Japanese.

Crops and Cultivation

The main vegetable crops grown were potatoes, pumpkins, vegetable marrows, long green cucumbers, watermelons, rockmelons, beans, lettuce, cabbages, cauliflowers, tomatoes, carrots, and parsnips. One or two gardens grew rhubarb and broad beans, but these were the exception rather than the rule. Their method of cultivation was to use a single furrow moulded-board plough and harrows, horse-drawn, together with hand tools, such as forks, three-pronged hoes, and ordinary hoes. By today's standards, these methods would be considered behind the times, but it should be remembered that these were the accepted methods in that period. Irrigation was by the flood system, with vegetables grown in long narrow beds, about one metre wide, with an irrigation channel on each side of the bed. The beds were worked up to a fine tilth using forks and hoes. The narrowness was specifically designed to allow the water to soak the bed and, simultaneously, allow the bed to be worked from both sides, thereby eliminating the need to stand on the actual cultivated area. There was very little plant disease and few insect pests in the early days. Their only controlling measure was a bluestone-and-lime spray and dusting with tobacco dust on cabbage and cauliflowers.

Fertilisers

Fertilisers, as such, were only used for the natural ones of blood and bone and bone dust. The main source of manure was either horse or sheep manure. The horse manure was purchased from agents in Sydney by the railway truckload. The amount in a rail wagon varied with the class of wagon. 'D' class wagons were of wooden construction with a capacity of 8 tons. 'S' class wagons were of steel construction and held 12 tons. Loads of manure usually weighed about 6 tons, and because the system of freight charged was based on the truck's capacity, irrespective of the amount contained therein, 'D' class wagons were preferred. The manure originated in Sydney's racing stables and livery stables. It was about half straw or wood shavings and half manure, the straw providing humus in the mixture. The sheep manure usually came from Flemington sale yards, from agents who cleaned out the manure, either from railway sheep vans or the yards of the sale yard. For some reason, the Chinamen preferred sheep manure for their tomatoes and horse manure for other vegetables.

The rail trucks came to Camden Railway Station, and horse and cart then carted the manure to the gardens. Costs would have been something in the order of price paid to the agent £2/10/0 railway freight £1/5/0 local carter to unload and deliver, 15/- or about £1 for about six cubic metres of manure and humus. Sheep manure was dearer, the difference being in the agent's price. With the decline in horse and sheep manure availability in the late 1930s, more artificial fertiliser was used.

Manure and vegetables were carried around the garden by manpower. The Chinaman used a two-carry-basket system, each about thirty inches in diameter and twenty-four inches deep. Each basket had two handles, from which a rope was looped over a pole, usually of hickory wood, carried on his shoulder. By straightening his back, the baskets were lifted off the ground and carried by a 'jogging' action rather than a straight-out walk. It was not unusual for them to carry up to 300 lbs for 300 to 400 yards.

Irrigation

One of the oldest and most experienced gardeners usually prepared the beds for irrigation. By very patient use of a type of harrow drawn by a draught horse, the Chinaman laid out his bed to make the best use of his flood irrigation. The 'Chinamen's Harrow' was like an overgrown pronged hoe – about four feet six inches wide, with prongs six inches apart. The operator held it in a vertical position, and by inclining the harrow forward or backward, it was amazing how quickly and evenly the soil could be spread. By the late 1930s, contract hoeing was used to a small degree, mainly to break up the ground initially, even cut irrigation drains, and chop up vegetable residue. In general, the Chinamen considered that rotary hoes produce a 'hard bottom' on garden beds. One of the earliest operators of rotary hoes was the late Bert Price.

The pumping of water from the river was mostly by a four-inch centrifugal pump, belt-driven from a 15 HP portable Marshall or sometimes a portable Ruston steam engine. There were five portable engines in use. A portable engine was, as its name implies, portable – being on four wheels. The two large wheels at the rear were on a fixed axle, and the two front wheels, about half the size of the rear ones, were on a movable turntable and were used for steering the engine when moving. One or two pairs of shafts could usually be fitted to the turntable at the front, and then four or six draught horses were harnessed to the engine, and the unit was moved to wherever it was required. Here, the engine would be set up as close to the river as possible, on a level patch of ground. If flood waters inundated the engine, little damage was done. Mainly it was a matter of washing out the bearings with kerosene, replacing the oil, checking the oil, checking the leather belt, lighting up the fire, raising steam, and it was ready to go again! The firebox was at the rear, then the boiler, and the smokestack in front of the boiler. The twin cylinders were mounted on top of the boiler, and the motive power produced by the two pistons drove both the flywheel and the main drive wheel. A six-inch-wide leather belt connected the drive wheel to the pump, which was as close as possible to the water source.

Coal was the basic fuel used for burning. It should be realised that coal was not being mined locally now, and it had to come from mines either on the South Coast or the Newcastle area. Again, the transport was by rail to Camden, thence horse and cart to the garden. Leslie Nixon, Albert Wood, Edward Hayter and William Taplin were local carters employed to cart the coal. Yee Lee used a slightly different system of steam boiler, engine and pump, a six-inch pump, and for many years, this was the largest pump on this section of the river. There were also certain technical features about this particular pump, which made it unusual.

In those days, there was no specialist irrigation pump serviceman to call upon. The Chinaman did his servicing. Similarly, there was no annual inspection of the 'Pressure Vessel' as is now made by the Department of Labour and Industry. In the whole period, there was no record of any boiler explosion. The Chinamen did call on the skills of such men as Joe Keane, a former blacksmith who had a motor garage adjacent to the Crown Hotel; Stan McKnight, engineer of the local milk factory; Billy Rapley, from Rideout's sawmill; and all played a part in keeping the steam engines working for the Chinamen.

Steam gave way to electricity and an odd petrol engine in the late 1930s because the steam engines were wearing out and parts were no longer available. Also, no engine man was required, releasing another man to work in the garden. Men like George Thornton, a motor mechanic, and Charlie Ricketts, an electrician, provided maintenance in this period. An interesting sidelight is that in the middle 1930s, Tong Hing Garden changed from their worn-out steam engine to a gas engine to drive their pump. The Camden Gas Company, whose manager was Frank Bean, piped gas from their town gas works in Mitchell Street about one mile across a road, paddocks, a dam, and through the market garden to the riverbank to provide a gas supply for the engine. One-inch galvanised pipes screwed together were used. The net result was that when the Chinamen started the gas engine, there was no gas left for town householders! This power source had a short life, and Tong Hing soon installed an electric pump.

Harvest Time

The Chinamen, at periods of maximum production, would employ local casual labour to help them dig their potatoes, pick their beans, or weed their lettuce. The prevailing system called for a fixed fee per bag when digging potatoes using a fork. During the depression of the late 1920s and early 1930s, some of the locals, whilst digging and bagging potatoes for the Chinamen, arrived at a 'scheme' where they added a tin of earth clods to each bag, thereby filling the bag more quickly, filling more bags per day and earning more money for the digger. The Chinamen soon discovered the ruse and promptly reduced the bag rate.

Potatoes were bagged in corn sacks, about forty inches by twenty-four inches, holding about 150 lbs, as were beans. Lettuce was packed in chaff bags, about forty-eight inches by thirty inches, not in boxes as at present. Up to the early 1930s, root vegetables like carrots, beetroot, and parsnips were not washed, being tied together in bunches with a strip of flax and sent to market with the dirt still on them. The Chinamen's idea was that washing and rubbing them to remove the dirt also removed the skin, and the root vegetables lost their freshness. In 1935, the Department of Agriculture insisted on facilities being provided to wash all root vegetables. This was usually a brick enclosure, plastered with cement to make it watertight, about eight feet by four feet and three feet deep, surrounded by a concrete slab. Vegetables were washed in a well and afterwards allowed to drain on the concrete.

Tomatoes were grown in long beds, with two rows to a bed, not staked or pruned, but allowed to grow on the ground. Flood irrigation sometimes made them dirty, but they were not washed, only wiped. To try to keep the fruit from being sunburnt or scalded, the Chinamen would scatter straw over the vines. The fruit was packed in wooden cases about thirty inches long, six inches wide and nine inches deep, with a centre wooden divider. Pumpkins, marrows, melons, cabbage, cauliflower, carrots, beetroot, and parsnips were packed on the truck in single units. Using the leaves or bunch tops packed under one another, it was amazing how firm a load could be packed on.

Marketing of Crops

In the late 1920s, all vegetables grown were sent to Sydney by train. The Chinaman or some local carter with his horse and cart would bring the vegetables to Camden railway, load them into a louvre van, class LV, and at 4.30 p.m. or 8 p.m., the train would take them away. At about midnight, the trucks would arrive at Darling Harbour goods yard in Sydney, where they would be unloaded and using horse and cart, the vegetables would be taken to the Sydney produce market. Having travelled to Sydney on the 8 p.m. train, the marketman would take over and sell the goods.

In the late 1920s, motor lorries commenced to pick up vegetables at the garden and deliver them straight to the Sydney market. With a couple of short breaks – one caused by a government act in the early 1930s requiring the use of the railway and one during the 1939-45 war to save petrol – this system prevailed to the end of the main era.

Carriers who come to mind include Messrs. Archie Tippetts, Albert and Allan Gander, Tom Holly, Jim Marden, Ern Richardson, Norman Richardson, Jim Bond, and the Bond Brothers. These are just a few of those who provided this service.

Figure 5 - Bond Brothers' truck taking Chinese produce to Sydney market, 1964 (Mrs Lyn Luscombe, nee Bond).

An important part of the local community

From about 1910 to the mid-1940s, the Chinese community living within the municipality of Camden was quite large and, indeed, close to the town area. The number was 75 to 80, almost all being males – only two or three Chinese women being among them. This number may not be considered sizeable by today's standards. Still, when one remembers the town had a population of less than 1000 in 1900 and about 2000 in 1940, it represents something in the vicinity of 4 percent to 5 percent of the total population. The 1921 Commonwealth Census shows that within the municipality of Camden, there were 970 males and 1036 females of all ages at this date.

If we look at the male figure and make allowances for the children under twenty-one years and remember the 'male baby boom' following World War I – the adult male population would not have exceeded 700, possibly less, and with upwards of eighty of these being Chinese – we have a situation where something like 10 to 11 percent of adult males was Chinese. A very sizeable group by any standard.

Furthermore, in those days, the municipality extended only from Wire Lane, adjacent to the Camden Valley Inn, to a point adjacent to Narellan Hotel; south-westerly to near Cawdor Methodist Church; westerly to the junction of Druitt's Lane and The Old Oaks Road; thence northerly to Sickles Bridge – this area of about four miles by three miles, including Elderslie, Camden, and the Carrington. Imagine around 2000 people in that area, of which eighty were adult Chinese.

Contribution to the Community

Another point concerns the financial contribution of Chinese gardeners. With the establishment of the Camden Municipal Council in 1889, the municipal rating system in use at that time was based on the 'assessed productive value' of the property. The valuers for making such assessments were two aldermen, elected out of the council for one year. For the first twenty years of its existence, the council was never too financial, but such was the economic influence of the system. I quote, "The finances of the Council were improving – the Chinese market gardeners paying big rents for gardens adjacent to the river so increased the rating value, and this was naturally reflected in the increased rates collected." In other words, this group enabled the general rates to be kept at a lower level.

Honour and Trust

The Chinese community was law-abiding. The men worked on the principle that all debts had to be paid by the Chinese New Year. This date varies from year to year but generally occurs in February-March. Because of this policy, their credit was always very good with the local tradesmen and shopkeepers. Bad debts by Chinamen were virtually unknown. If the season had been bad or there had been floods, the local debts were always paid, even if they had to take a loan from Chinese merchants to meet the same. Interest charges on these loans were usually high. Occasionally an odd Chinaman would be charged for drunkenness. Early in the century, there was an odd charge of opium smoking.

The community's greatest numerical strength was between 1910 and 1942, and during this period, three group leaders stood above the others. Willie [Chung] of Sun Chong Kee (Sheil's), Charlie of Yee Lee's garden, and Wong of Tong Hing's (Davies') garden were leaders of their community and gentlemen of the highest order in every respect of the word. They were very highly respected by all who knew them. Any tradesman or shopkeeper who had any problem with any member of the

Chinese community had only to speak to one of these gentlemen, and the problem ceased to exist. Willie [Chung] was the eldest by far of the three leaders, and with traditional Chinese respect for age, his countrymen held him in the highest esteem. [Willie Chung's story is in chapter 11.]

Figure 6 - Willie Chung, Immigration papers, 1946 (National Archives of Australia ST84/1, C1946/5927, Page 97/97).

Funny Names and Ways

Locals generally were not interested in something other than the rather colourful Chinese names of the groups. At Nesbitt's, there was Sing Mo; at Thurn's, there was Hop Chong; and at Whiteman's, there was Young Lee, sometimes known as Yong Lee. Later still, Hop Chong Company took over the Whiteman's Garden. This group name was known in the district for something like 60 years. The last Chinese occupant, Biu Wong, took over in about 1970. At Sheil's, there was Sun Chong Kee for thirty years; at Davies', there was Tong Hing for fifty years.

The small family unit at Elderslie disposed of their produce generally by peddling it around the town, first by horse and cart and later by a 'T' model Ford truck. He was known locally as Charlie the Chinaman, or sometimes as 'Big Charlie' to distinguish him from 'Little Charlie', who was the boss of Yee Lee. As previously stated, most were known by their anglicised nickname.

In the early days, any Chinaman who died was buried in the local general cemetery without a Christian service. Still, with the introduction of the motor hearse, burials were at the Chinese section of Rookwood Cemetery. Following the old Chinese custom, their bones would be exhumed (after the legal lapse of time) and returned to China. There was no local joss house, the closest being in Sydney.

Working seven days per week, they worked 363 days per year. One of their non-working days was Chinese New Year, a celebration day, and strangely, they did not work on Christmas Day. A few days before Christmas, they would go around their business acquaintances, such as storekeepers, butchers, railway station masters, police sergeants, and carters. If the season had been good, they would hand out gifts of preserved ginger in crockery jars and dried lychee nuts.

2 • Farming Practices by the Chinese Community of Camden

Figure 7 - Camden view looking along John Street from the steeple of St. John's. The house of the Davies' market gardeners (Tong Hing garden) is visible in the far distance near the river. The site of the vegetable crops is currently part of the Miss Llewella Davies Walkway.

Before World War I, efforts were made to prevent them from working on Sundays, but this was unsuccessful. If the police received a complaint about Sunday working, the police were obliged to call and order the Chinamen to cease work. This was unworkable for the Chinamen professed 'not to understand', so no action could be taken. The regulation was just not capable of being enforced.

Usually, they did not participate in any of the local sporting organisations. To the best of present knowledge, only one from Tong Hing's (or Davies') garden joined a sporting group. He played football for one season with Camden Football Club. He still had to do his share of gardening, even though he was away on Sunday afternoon playing football. It was not unusual to see him ploughing, with horses, with a hurricane lantern after dark and quite late at night so he could have time off to play sport. This Chinaman was of a younger generation and was most certainly born in Australia.

Bibles

About 1930, the Church of England rector of the day, the Rev T.G. Paul, obtained several New Testaments from the British & Foreign Bible Society, printed in one of the numerous Cantonese dialects. These he left at the Camden Hospital to provide reading matter for any Chinese admitted to the hospital.

The Authorities

From time to time customs and immigration officials, assisted by police, would make 'raids' on gardens, looking for illegal immigrants. One remembers a particular raid in the early 1930s when several illegal immigrants ran off and hid in the river, with only their heads out of the water, hiding in water reeds. They were in the water, the officers on the bank waiting. The illegals would not come out; the officers would not enter the water. Eventually, however, they came out and, in due course, were deported.

During the Sino-Japanese War, they were intensely loyal to the land of their birth, sending home money for the war effort. One remembers one Chinaman leaving for home, and his one aim was to take with him sufficient money to buy himself a gun to use against the Japanese.

It should be remembered that the Chinese mentioned had entered Australia before about 1900, and whilst they were free to return from China, no new ones were admitted into the country to take their place. They worked hard, from daylight to dark, under what one now calls 'coolie conditions', and the young ones born in our country wanted a freer existence with easier and better-paid jobs.

One can recall the older men packing up to go home to China. Indeed, they would go around their acquaintance before they left and frankly state they were going home to end their days. Up to about the age of fifty-five years, they would go home, stop for a few weeks, and then come back, but at about sixty years, they went and never returned. In other words, no one took their place. So old age was one cause of their passing.

End of an Era

For some unknown reason, probably old age, the Chinamen did not appear to be able to cope with modern vegetable-growing techniques. Generally, they appeared unable to handle a tractor; they did not grow tomatoes on stakes; they grew only vegetables; they did not appear to be able to adapt to modern spraying and disease control, and so on. The vegetable growers of the district are now of Italian and Yugoslav descent.

And so, into history has passed a phase of our community, and I am sorry to say it is now a little-known phase! All sections and phases of a community's life are vital in that community's history and our nation's history. Let us not forget that 5 per cent of any community is a sizeable number and that the Chinese of this town equalled, and at times exceeded, that number.

Editor's note: Richard Nixon collected for the Camden Museum several items that remember the Chinese market gardeners in Camden, and he encouraged John Wrigley to record the following stories of the objects.

2 • Farming Practices by the Chinese Community of Camden

Figure 8 - Dianne Holmquist nee Wong (dark blue jumper) and Holmquist family visiting Camden Museum, 10 June 2019 (photo by Roslyn Tildsley).

3 • Market Gardeners' Farming Artefacts in Camden Museum

John Wrigley

Almost all the items at Camden Museum have been donated by residents. After the Museum was established in 1970, several of the members of the Camden Historical Society were particularly interested in the Chinese market gardeners and set up a bay to remember them. Having a bay in the Museum encouraged people to donate items, some of which were dug up years after the Chinese gardeners had gone. The Museum honours the contribution of the Chinese market gardeners to farming and settlement in the Camden area.

Watering Cans

On display, the Museum has two large, galvanised iron watering cans used by Chinese market gardeners in Elderslie, part of Camden on the eastern side of the Nepean.

Figure 9 - Watering cans at Camden Museum. Items 2012.26 and 27 (John Wrigley, 2014).

3 • Market Gardeners' Farming Artefacts in Camden Museum

The watering cans were used on land owned by the Thurn family under the present Macarthur Bridge, Camden. The local donor is Mr. Bruno Carmagnola. The donor's parents, Silvio and Maria Carmagnola, were Italian migrants who came to Camden after the Second World War. They lived and operated their market garden and orchard on the property on Macarthur Road, next door to the Chinese market gardeners.

The watering cans date from about the 1940s and 1950s, but there have been cooperatives of Chinese market gardeners in Camden for much longer.

Bruno Carmagnola picked up the watering cans after the Chinese market gardeners had left the area. He stated in 2014 that he found the watering cans on the Thurns' farm in the old shed previously used by the Chinese gardeners near the Nepean River, away from the Thurn house close to Macarthur Road. He said that the market gardeners could carry the buckets on their shoulders on each side of a wooden yoke and walk between two rows of vegetables, watering the row on each side simultaneously. He said the buckets were heavy even without the water in them.

On display with the watering cans is a carrying stick or yoke, 155 cm in length, which Chinese market gardeners use on their shoulders to carry two watering cans at once. Richard Nixon donated it, and it represents the hard physical work of watering market gardens by hand. Often, the Chinese market gardeners ended their lives with bent backs.

The Chinese market gardeners were one of the few non-Anglo cultures to be part of Camden's history, so these two watering cans are significant in their association with the Chinese who came and worked in Camden. Hanging from the ceiling, the watering cans significantly contribute to the Chinese display in Camden. They may be the only remaining examples of the heavy watering cans used in Camden during the era of the Chinese gardeners.

Objects like these watering cans are so humble and ordinary that they are not often kept, but they remind us of a Chinese community who used to work on the fringe of an Australian country town. Together with the other objects in the display about the Chinese market gardeners, the watering cans help to interpret the history and cultural contribution of the Chinese to the country town where they lived and worked.

3 • Market Gardeners' Farming Artefacts in Camden Museum

Steel Rake

Two men pulled this by hand or, later, by horse. Bruno Carmagnola donated it after the Carmagnola family kept two rakes on their orchard, on the property adjacent to where the Chinese market gardeners had worked. Each rake or harrow would have had a pole-shaped wooden handle fitted through the two top eyelets for pulling it across the gardens. The Chinese market gardeners used the rake to break up the soil before planting crops. They represent the physical hardship experienced by the Chinese market gardeners.

Figure 10 - Two of the steel rakes which were used by the gardeners. Item 2012.25.

3 • Market Gardeners' Farming Artefacts in Camden Museum

Davies' Rent Receipt Book

The rent book is a small green stapled notebook with hand-written entries. It was started in 1949 by Mary Fabert Davies and taken over by her daughter, Miss Llewella Davies, after 1954. The last entry was in 1963. The book recorded regular rental payments for the use of Tong Hing Garden (known as Davies' garden).

Each payment was stamped to show that an official stamp duty payment had been made for that amount. The stamps would have been purchased at Camden Post Office. The leader of the market gardener cooperative, Tong Hing, paid regular visits to the house of Miss Davies in Exeter Street to make the payments. The fact that the book was kept long after it was used shows that the relationship between the Davies family and the Chinese cooperative was of mutual benefit and the contribution of the gardeners was valued.

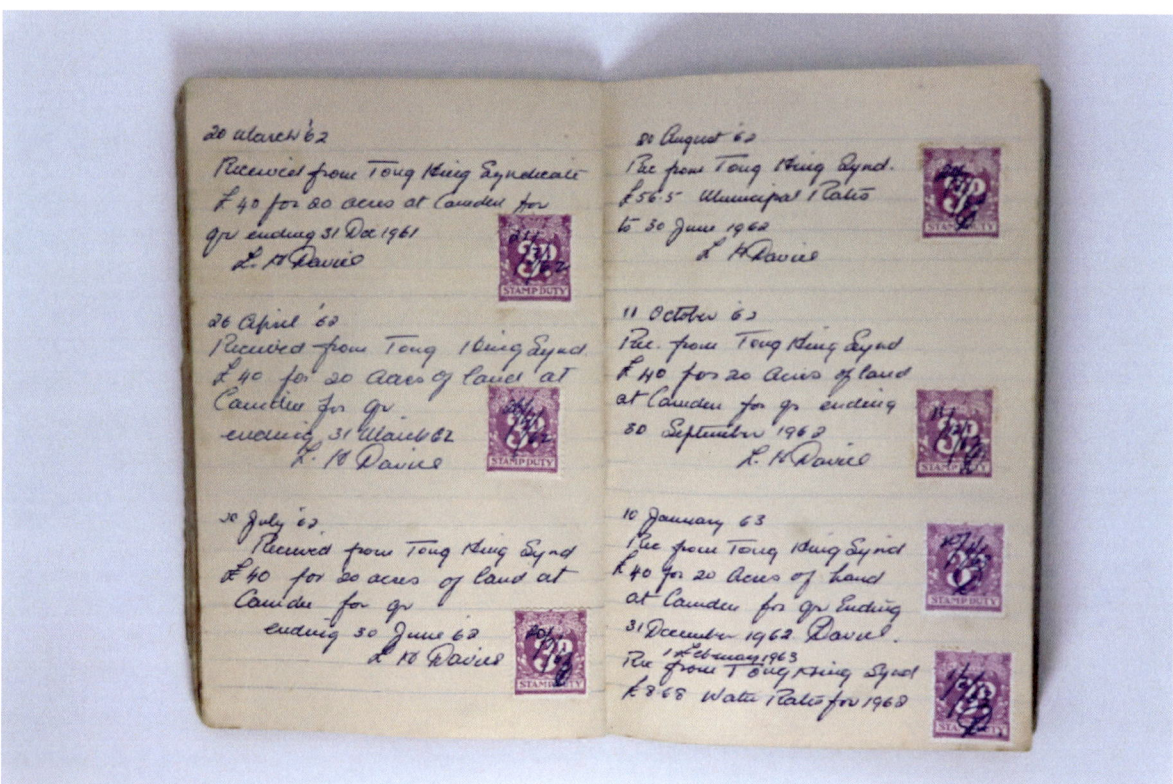

Figure 11 - Miss Davies' Tong Hing Rent Receipt Book for 1962-3. Item 2022.16 (Anne McIntosh).

Davies' Lease

The original document (39 x 26 cm) of the Tong Hing Lease in Exeter Street, Camden, is the formal Memorandum of Lease between Mary Fabert Davies and the Chinese Sun Tong Hing Company of Camden market gardeners. It was signed by seven lessee members of the company or syndicate: Foon Kee, Ho Jo Ling, Jack Young, Law Sang, Ah See, Solomon Fungnam and Yen Hung Pan and witnessed by local solicitor R.A.C. Adams on 31 October 1949. The lease with the Davies family was renewed from 1949 to the 1960s. It is very interesting to see the signatures of the gardeners.

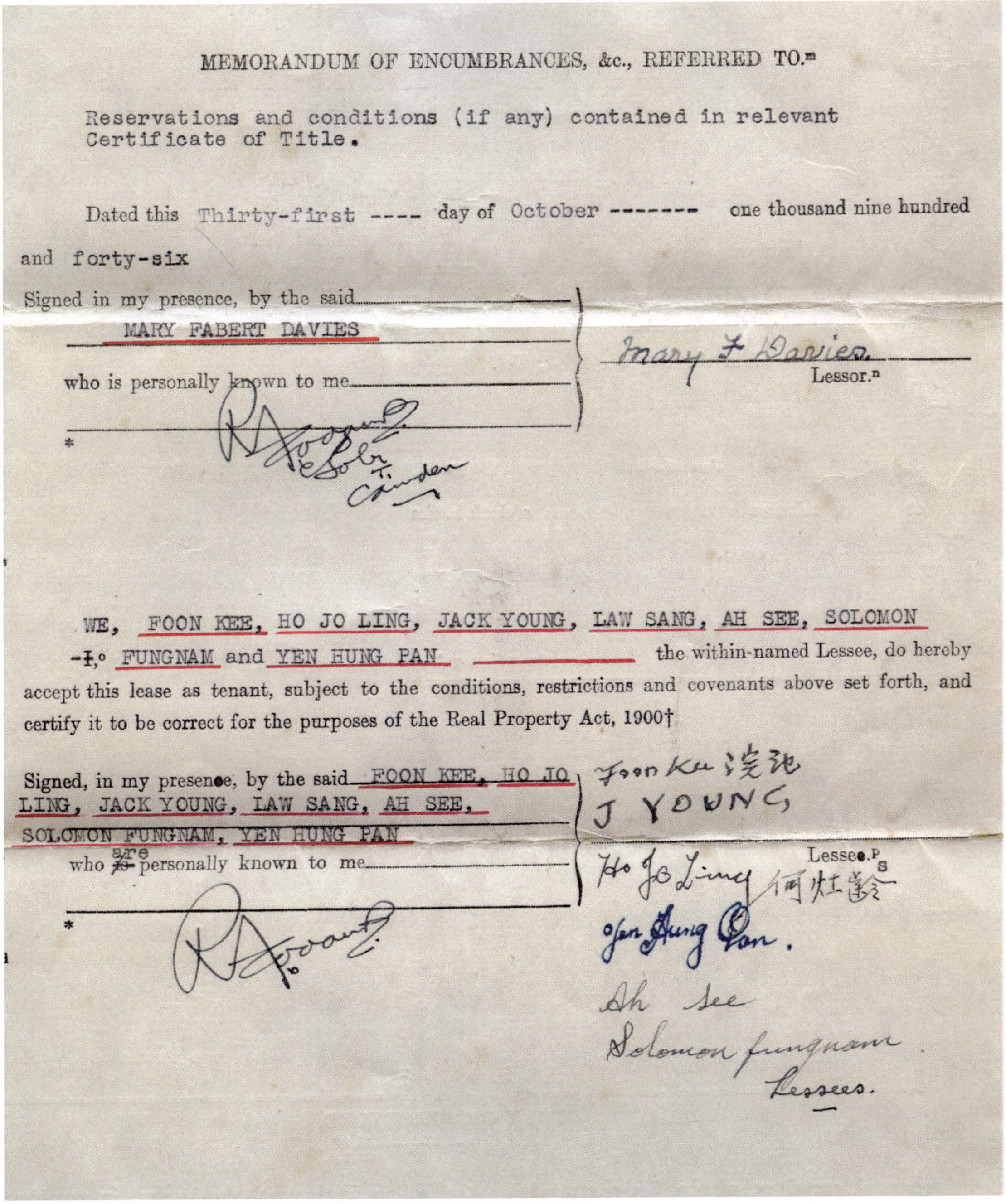

MEMORANDUM OF ENCUMBRANCES, &c., REFERRED TO.

Reservations and conditions (if any) contained in relevant Certificate of Title.

Dated this Thirty-first ---- day of October ------- one thousand nine hundred and forty-six

Signed in my presence, by the said MARY FABERT DAVIES who is personally known to me

Mary F Davies. Lessor.

WE, FOON KEE, HO JO LING, JACK YOUNG, LAW SANG, AH SEE, SOLOMON FUNGNAM and YEN HUNG PAN the within-named Lessees, do hereby accept this lease as tenant, subject to the conditions, restrictions and covenants above set forth, and certify it to be correct for the purposes of the Real Property Act, 1900†

Signed, in my presence, by the said FOON KEE, HO JO LING, JACK YOUNG, LAW SANG, AH SEE, SOLOMON FUNGNAM, YEN HUNG PAN who are personally known to me

Foon Kee
J YOUNG
Ho Jo Ling 何灶齡
Yen Hung Pan.
Ah See
Solomon fungnam
Lessees.

Lessees.

Figure 12 - Davies' Market Garden lease, 1946. p. 2 (Camden Museum).

3 • Market Gardeners' Farming Artefacts in Camden Museum

'Chinaman's Garden, Camden NSW'

This sketch of a Chinese man in a 'coolie' hat carrying two objects on a water bank levee comes from the book *Douglas Annand – Drawings and Paintings*. Douglas Annand (1903-1976) was a graphic designer and artist. In 1941-44, he was a camouflage artist with the Royal Australian Air Force. [22] While he was stationed at Camden, he sketched the Chinese market gardener. This is the only known artwork showing one of Camden's Chinese market gardeners at work.

Figure 13 - 'Chinaman's Garden, Camden NSW', sketch by Douglas Annand, from the book Drawings and Paintings 1944 (Item 2022.18.).

---

[22] Australian Dictionary of Biography, https://adb.anu.edu.au/biography/annand-douglas-shenton-9369

4 • Interaction with Camden Community

Julie Wrigley

In Camden, as in many NSW towns, Chinese and non-Chinese went about their own business. Non-Chinese residents marvelled at the diligence of the Chinese in raising vegetables on the river flats. Still, they did not have the skills or inclination to work like that and were not interested in competing for the jobs the Chinese undertook. There was very little interaction for various reasons, including differences in language and culture. Market gardeners kept to themselves, working in their own cooperative and most non-Chinese residents in Camden were content to keep their distance and to see the Chinese gardeners as outsiders. However, several groups of local people had frequent contact with the Chinese market gardeners, and their relationship was closer.

Carriers

Richard Nixon's father, Leslie Nixon, was a carrier who carted vegetables and worked closely with the Chinese. In October 1996, Richard gave a talk to the Camden Historical Society, which was later transcribed from a tape made at the lecture. The topic was Chinese Gardens in Camden. Richard talked about the carriers who worked with the market gardeners.

The Picton Post (18 May 1944) referred to the interaction between the Chinese and the carriers. When one of the local carriers, James Adam Bond, died, two Chinese market gardeners attended the funeral as a mark of respect:

> A further touching tribute was paid on the occasion by two Chinamen who, at the conclusion of the service, knelt at the foot of the grave and went through their simple ceremony of farewell.

Owners who Leased Land to the Chinese

Chinese leaders of the cooperatives visited the landowners' homes regularly to pay the lease rent. Residents reminisced about contact with the Chinese market gardeners and how honest the Chinese were in their dealings. Mary Locke grew up at the property *Caernarvon*, which looked down on Sheil's garden on the river flats. Mary Locke remembered the Chinese men who worked hard supporting families back in China, 'I have nothing but praise for them. They were beautiful whenever we went down there.'[23]

The leases along the river were advantageous to both Chinese and non-Chinese residents. The Chinese could sell their produce to the city but also supplied residents with high-quality, locally-grown vegetables.

Bakers

Frank Stuckey, the local baker, said the Chinese 'were very good people to deal with'. He wrote an account of the Stuckey bakery in Camden, titled *Our Daily Bread – The Story of Stuckey Brothers,*

---

[23] https://camden.spydus.com/cgi-bin/spydus.exe/ENQ/WPAC/BIBENQ?SETLVL=&BRN=120116 Camden Voices oral history interview, 2012.

4 • Interaction with Camden Community

Bakers and Pastrycooks of Camden, page 30. Referring to Camden in the 1920s and 1930s, Frank Stuckey wrote,

> Way back in the early days, the Chinamen from the vegetable gardens used to come to the shop early in the morning for hot bread. In those days there were six gardens. The names of the market gardeners were Yee Lee, Young Lee, Tong Hing, Hop Chong, Sing Mo and Sun Chong Kee. One garden was either side of the main road towards the bridge [Cowpasture Bridge], another was near the weir on the Sydney side of the river, one garden on what is now Miss Davies' property near Macquarie Grove Bridge, and two were on the southern side of the present Macarthur Bridge. One property is now farmed by the Carmagnolas and another property would now be Mr. AA Tegel's property. At the Carmagnolas' garden we used to deliver the bread.

At a historical society meeting in the 1970s, Frank Stuckey said:

> I worked on just about all those gardens – about four out of the six of them, but more particularly where they're playing [softball] where the ladies are playing down there – where they're building the ablution block…Yee Lee's as it was, I used to pick a lot of beans there, any time after school. In the Christmas holidays we had more time and we'd pick down there of a morning till it got too hot and then we'd go for a swim and we'd come back and pick some more, but we made very good pocket money down there and I remember Richard Nixon's dad who was associated with them, and, of course, they used to come up to our shop for their bread, we never delivered. Oh, we did deliver up the river there, but most of them came up to the shop for their own bread and they loved French bread and the cook would come up, he was Number Three or something, he'd come up and he'd be waiting for the bread to come out and he'd get whatever he wanted and go home. They were very good people to deal with. [24]

Children

Children earned money from the market gardeners, mainly from Hop Chong Garden. In 2015, resident Richard Cornhill told John Wrigley that in the 1950s, Richard and his friends used to catch carp fish and sell them to the Chinese market gardeners for ten shillings a pair. The gardeners used the carp to clear the water in their concrete water tanks and small dams. The boys also used to sell the market gardeners brown table pigeons for eating, for ten shillings a pair. Richard said this earned them 'serious pocket money'. [25]

Flood Rescuers

Camden had many floods, and the Chinese market gardeners needed to be rescued in the flood boat when they reluctantly left their homes. The Chinese were taken to the Town Hall, formerly the School of Arts, now the Library/Museum complex, and given food and shelter until the flood went

---

[24] Formal thanks by Frank Stuckey to a talk given by Richard Nixon to the Camden Historical Society, 9 October 1996: 'Chinese Gardens in Camden'. Transcribed by Dianne Matterson in 2016 (CHS archive box: Chinese market gardeners).

[25] Handwritten note to John Wrigley by Richard Cornhill, (CHS archive box: Chinese market gardeners.)

down. In the newspapers, there were frequent reports of floods and rescues of Chinese market gardeners. 'Camden in flood' (*Camden News* 19 January 1911); 'Chinese marooned, excitement at Camden' (*The Maitland Daily Mercury* 12 May 1925); 'Camden isolated by flood waters' (*Sydney Morning Herald* 21 May 1943); 'Timely rescue by police' (*The Picton Post* 23 June 1949); 'Nepean flood threatens Camden area' (*The Sun* 20 March 1950.)

Sydney's *Daily Telegraph* (21 May 1943) had a heading 'Chinese Flooded out of Garden' with a photo of the Chinese market gardeners Ah Ling and Lai Wah who were 'evacuated from their flooded riverside gardens spending the night in the Camden Town Hall'. Another photo on the same page showed Constable A.E. Neale 'handing out food to some of the gardeners after helping to evacuate them from their garden by rowing boat.'

There were Camden floods in 1949 and 1950. *The Daily Telegraph* Sydney (7 April 1950) stated:

> 'Last night Camden police evacuated eighty Chinese market gardeners and gave them temporary accommodation in the Agricultural Society Hall'.

Figure 14 - The Chinese were rescued and accommodated in the Town Hall (*Daily Telegraph* 23 May 1943).

4 • Interaction with Camden Community

Figure 15 – A rare photo of Camden Chinese gardeners washed out by a flood in 1925 (Camden Museum).

Engineers

Local engineers John and David Southwell, who grew up in Camden near the gardens, recalled the 1940s and 1950s. John said the Chinese gardeners kept to themselves, and there was no animosity. Local Camden people liked the men working on the gardens along the Nepean River, and they were thought to be an asset to the town. John said the Chinese had a deep well where they kept carp fish for eating. They had carp in their ponds and would walk down a ramp into these ponds to fill their watering cans and hand water their gardens. Their pumps were high-volume/low-pressure pumps, so hand watering was necessary. Each garden had a horse and sledge for transporting produce and goods around the town. They used to bring the sledge into the Southwells' engineering workshop to repair the sledge and chains. [26]

Camden Hospital

In the 1940s and 1950s, the Chinese regularly donated to Camden District Hospital funds, and their donations were recorded in the local paper. There were donations for the Hospital Equipment Fund from Yong Lee garden, Yee Lee garden, Sun Hop Chong Garden, and Tong Hing Garden. (*Camden News* 25 January 1951). The Chinese gardeners had their own medicines, but they valued the hospital.

Chinese and non-Chinese lived side by side in the quiet country town. The two communities kept their distance, but older residents all spoke of the Chinese with respect. Two Camden residents, Miriam Dunn and Joy Riley, have written the following recollections of the Chinese market gardeners.

---

[26] Notes from Mr John Southwell OAM in 2015. (CHS archive box: Chinese market gardeners).

5 • Memories of the Chinese Market Gardeners

Julie Wrigley

Miriam Dunn grew up on the Thurn family property on the Nepean River flats at Elderslie under the Macarthur Bridge. Their neighbours, the Nesbitts, were located about 500 metres south of the present Macarthur Bridge. Miriam wrote these notes for the Museum in 2015. Joy Riley wrote her notes in 2016.

Miriam Dunn (née Thurn)

I was very young when the Chinese market gardeners worked on the family property. Being a girl, I was not allowed to go to their place unless I was with my father or older brothers. The Chinese kept to themselves and were always very polite. My parents always took them milk, butter, cakes and biscuits, and fruit when we had some from the orchard. Sometimes we took them a chicken and eggs. In return, they always gave us vegetables.

When we had a flood, they would always stay at our place. We had a room near the house they stayed in and left their things in until they could return to their shack (which is all you could call it. It never had a wooden floor, in parts of it just dirt.) They were always happy and never had any arguments.

While the floods were on, they visited Sydney to visit family or friends. Richardson's truck used to take their produce to the markets several times a week. The butcher used to deliver meat to them, and one of the grocery stores used to deliver groceries. When the boss went to Sydney to collect the money to sell their produce, he often brought rice and other things they could not get at the local grocery store.

I cannot remember when they left our property. When they left, they moved to a property near the milk depot and later an Italian restaurant [at three Argyle Street].

The only drama I can remember was a big flood on Nesbitt's property (leased by San Yeck garden). The Chinese were evacuated. They used to stay on the verandah at Nesbitt's house (Galvin's Cottage). At the time, no one lived in the house, it was used only in holidays. The Chinese decided to take no notice of the authorities who told them to evacuate, and when the flood started to go down, they headed back home. However, there was extra water to come down from Robertson as it still rained up there.

Some people with boats tried to rescue the Chinese, but the current was too strong and put the boat into the fences. It was getting dark by this time, and the Chinese were seen on the roof flashing their torches. Eventually, someone with a boat from Camden could get through the current and rescue them.

5 • Memories of the Chinese Market Gardeners

Figure 16 - The Chinese garden sheds were flooded in 1943 (Mrs Lyn Luscombe. nee Bond).

Joy Riley (née Dunk)

My father, Jack Dunk, ploughed the Chinese gardens of Miss Llewella Davies at Macquarie Grove, next to the sewage works, known to the locals ironically as the 'Rose Bowl'. (Figures 16 and 17).

In 1949 our family purchased a fruit and vegetable and mixed business at 152 Argyle Street – now a butcher and next door to Blooms Chemist. Our family was very friendly with the Chinese men as we purchased a lot of their produce. Sunday evening, Bond Brothers would load up cauliflowers, cabbages, and other vegetables for the Sydney markets. Some of the Chinese gardeners would travel with the Bond Brothers to Haymarket and come home to Camden on Monday mornings with my father after he had gone to the markets to buy stock for our shop. We also bought vegetables from the other gardens in the area.

When we had a flood, and the gardens were flooded, the men would not leave unless the Bond Brothers or dad would come and pick them up. At Christmas time, Dad would go down to the Crown Hotel and buy draught beer and have it put into one-gallon jars to share with the men from Miss Davies' property, although he did not usually drink.

After the war I remember the Chinese men walking down John Street in single file. They would have come from Thurn's or Nesbitt's gardens on the other side of the river in Elderslie, over the bridge at 'Little Sandy' at the bottom of Chellaston Street, and up through the rectory paddock, over the stile in Menangle Road. They all wore loose clothing and had a single plait of hair. I do not know where they were going.

5 • Memories of the Chinese Market Gardeners

Figure 17 - Chinese Gardens, Grove Road, Camden, near the Macquarie Grove treatment works. Miss Davies' paddocks in the background, Jack Dunk driving the tractor, Stan Rofe watching (Joy Riley).

Figure 18 - Jack Dunk on the tractor, preparing the soil for Chinese market gardeners, Stan Rofe standing (Joy Riley).

5 • Memories of the Chinese Market Gardeners

I remember two of the men from Macquarie Grove. One was named Foon [Foon Kee Pan]. He was the Boss and was always well-dressed and polite. He died of cancer in the 1950s. He gave the impression of being well-educated and a professional man. The other one was named Jo [Ho Jo Ling]. He was a very big man. He went back to China to see his wife, who was still in Red China.

One trip he brought back a gift for me, a beaded evening bag, which I still have. He also gave me a pair of slippers, but I was unable to wear the slippers as they were too small. My mother liked ginger and the Chinese men would give her the stone jars of ginger.

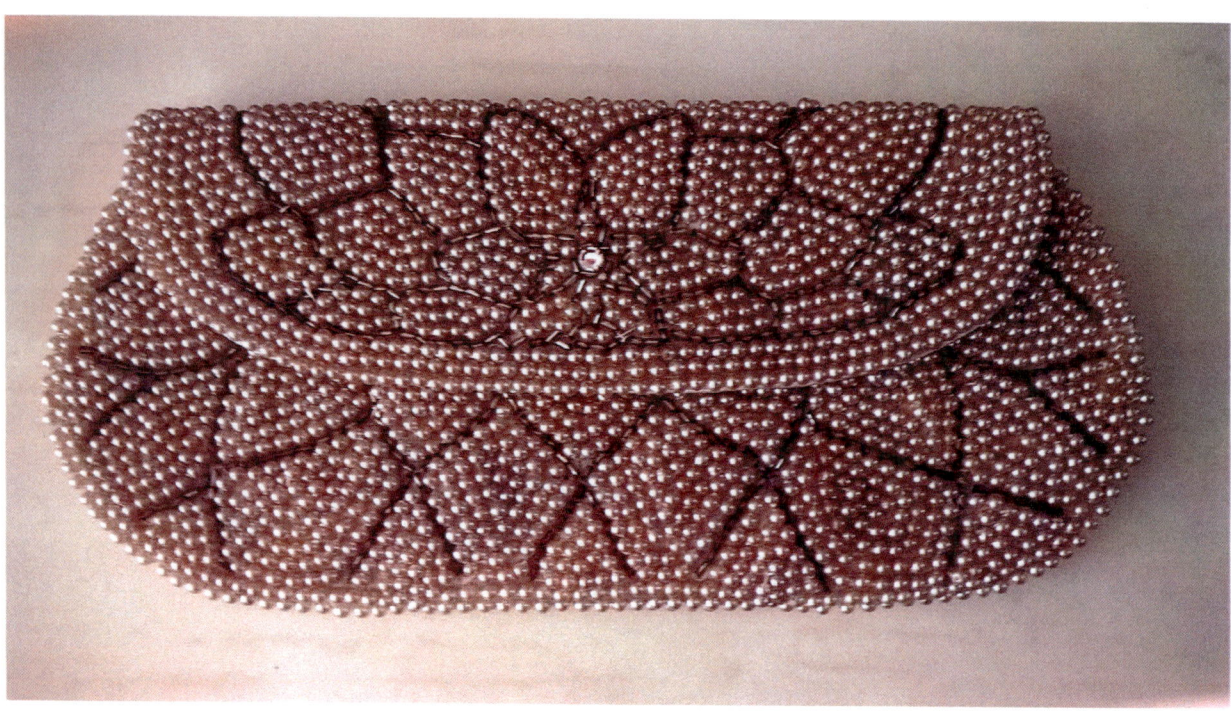

Figure 19 - Ho Jo Ling gave this beaded bag to Joy Riley when she was a girl.

6 • George Lee

The First-Known Camden Chinese Gardener

Julie Wrigley

George Lee, who leased land in Elderslie at the end of the nineteenth century, was the first Camden Chinese market gardener referred to in the *Camden News* (20 July 1899), which had a record of tenders for land to be leased:

> In consequence of several applications for Elderslie Church lands having been sent after the tenders were closed, the Archbishop has decided to call for fresh tenders. No tenders will be entertained that are sent after the 29th ins. [July, 1899]. The tender of Mr. George Lee, Chinese gardener at £4 per annum for lots 25, 26, 27, village of Elderslie, has been accepted.

A copy of this ten-year indenture agreement is in the Camden Museum files. [27] George Lee was eligible to sign the indenture as a step to own the land because he had been naturalised in 1884 [28] when the states, not the Commonwealth, arranged naturalisation.

The *Camden News* (30 June 1904) shows that George Lee offered some of his land for sale, but it was not sold:

> Mr. R.H. Inglis, auctioneer, offered at his Camden sale yards on Tuesday two allotments of land at Elderslie, the property of Mr. G. Lee, also a lease for twenty years of allotment adjoining the property, with a market garden in working order. Biddings started at £80, quickly went to £99 at which bid the property was passed in.

The land, or part of it, was sold a year later. The *Camden News* (14 December 1905) recorded the property sale:

> On Tuesday last, the 12th inst., Mr. W. Larkin (R.H. Inglis) submitted a market garden property at Elderslie, Camden, owned by Mr. George Lee, comprising an area of two roods 27 perches, having frontage to the main Camden road [Cowpastures Road], highly cultivated, water laid on, cottage, etc. On the property there was a lease rental of £20 per annum from Chinese workmen. The first bid was £50, which went quickly to £65, at £66 the hammer fell to the bid of Mr. George Baxter of Mount Hunter.

The NSW Land Registry Services archive records George Lee of Camden 'gardener and

---

[27] The Indenture between the Archbishop of Sydney and George Lee (Chinese gardener of Elderslie) was for ten years from 1899 to 1909 at £4 p.a. with half yearly payments in advance, for 2 acres of land, being allotments 25, 26 and 27 in Elderslie between Cowpastures Road and Harrington Street.

[28] George Lee naturalised 30 April 1884, [4/1208] p. 92. Reel no 133. https://www.records.nsw.gov.au/archives/collections-and-research/guides-and-indexes/node/1481

6 • George Lee – The First-Known Camden Chinese Gardner

greengrocer' selling, for £66, lots 24 and 28 to George Baxter, a farmer of Mount Hunter. [29]

In the 1901 Census, George Lee is recorded living in the same dwelling as others, who are listed as 'Chinese, can't speak English' [30]. In the Electoral Roll for 1901-1902 and for 1903-1904, George Lee of Elderslie is listed as a person entitled to vote as a Lessee with 'House and Land' at Elderslie. [31]

Prize-Winning Vegetables

It seems George Lee was hard-working and generous. *The Sydney Morning Herald* (21 March 1901) recorded the fifteenth annual Camden Show results: 'G. Lee won the prize for the vegetable collection of the district.' In the *Camden News* (24 April 1902), the executive committee of the Camden Cottage Hospital announced with pleasure that they had received from George Lee of Elderslie 'the handsome donation of £2/2s, the amount of his prize money awarded in the vegetable section of the Camden Show towards the funds of this noble institution.'. This was followed by an article in 1902 with a report from the Camden Cottage Hospital, 'A letter, conveying the thanks of the board, is to be sent to that gentleman for his donation.' (*Camden News* 15 May 1903).

The account of the seventeenth Camden Show in the *Camden News* (26 March 1903) gives a clear picture of George Lee and his personality.

> The interest in the proceedings on the ground was considerably heightened by the presence of His Excellency, Sir Harry Rawson, [Governor of NSW] Lady Rawson, and Mrs. Macarthur-Onslow and party, on both the second and third days. The genial disposition of the Governor and the keen interest he took in the show proceedings, mixing freely with the people and conversing with them, made him popular. An amusing incident occurred shortly after the Governor had declared the show open. George Lee, the popular Chinese gardener of the district, who, for the last two years had given the whole of his winning prizes at the Camden Show to the Camden Cottage Hospital, presented himself to the Governor with hat off and a polite bow, "Good day, Mr. Governor, gim me your hand [32]." His Excellency shook the extended hand with a smile, and thus won the heart of the people in one bound.

The *Camden News* (26 March 1903) listed the results of the Vegetable Section of the Camden Show. George Lee won prizes for the Best Collection of Vegetables grown in the district of Camden, best collection of onions, turnips, squashes, Ironbark pumpkins, lettuce, celery, beetroot, long radishes, turnip radishes and peas pod. The variety shows how successful George Lee was as a market gardener. He also won prizes in the Camden Show in 1905. (*Camden News* 23 March 1905.)

After 1905, there are no newspaper records of George Lee of Elderslie. However, archival records reveal George Lee's Certificate of Domicile dated 16 February 1905. [33] The certificate has quite

---

[29] NSW Land Registry Services Book 795, p. 551.
[30] 1901 Census, Elderslie and Narellan. Stephen and Christine Robinson, 2000, from State Records NSW.
[31] Electoral Roll, Municipal List for the Municipality of Camden, 1901-1902, p133; 1903-1904 p.171.
[32] Stereotype about the way Chinese spoke English contributed to Chinese being seen as 'other'.
[33] National Archives of Australia, George Lee: ST84/1, 1905/61-70. After the introduction of the White Australia Policy in 1901, Chinese travellers used Certificates of Domicile from 1901 to 1905.

6 • George Lee – The First-Known Camden Chinese Gardner

demeaning details: *Height: 5 feet eight inches in boots. Scar on left side of head above temple. Long scar on the inside of the left leg near shin.* The reverse page of the certificate has the impression of George Lee's left hand.

The Commonwealth authorities and customs officers were determined to follow the Immigration Restriction Act of 1901 to the letter.

George Lee made his trip to China in December 1906 on the *Eastern*. When he returned from China on the ship *Empire,* his Certificate of Exemption from the Dictation Test is dated 25 January 1909. The certificate is marked 'Returned from China. Naturalised'. [34] The government restrictions and discrimination contrast with the local paper, which described George Lee as 'the popular Chinese gardener of the district.'

Richard Nixon writes in his article on 'The Chinese Community of Camden' that the family unit was located on the Hume Highway at Elderslie, and it seems to be the same George Lee who leased land there in 1899. Nixon's map in the archives at the Camden Museum shows the location in Elderslie facing onto the Camden Valley Way. His map is marked 'Georgie the Chinaman 1920s'. Rookwood Cemetery has a burial record for Georgie Lee, aged 62. [35]

Figure 20 - George Lee, 1910 extracted from his Certificate Exempting from Dictation Test (National Archives of Australia: ST84/1, 1910/44/1-10. Page 9).

---

[34] George Lee, Certificate Exempting from Dictation Test, NAA: ST84/1, 1909/13/21-30.
[35] Rookwood, Georgie Lee, Chinese 3 Zone H 287. Allotment 287, Chinese Section, date of interment 28 December 1928.

6 • George Lee – The First-Known Camden Chinese Gardner

The Chinese market gardener George Lee was a colourful character who significantly contributed to the Camden district in the early twentieth century. His life story shows conditions in the early days of the Chinese market gardeners in Camden.

Figure 21 - George Lee's Certificate Exempting from Dictation Test, 1910 (National Archives of Australia: ST84/1, 1910/44/1-10. Page 9).

7 • Chun Yuen (1870-1921)

Accidental Death on the Road

Julie Wrigley

The Chinese market gardeners appeared in newspapers only when there was a flood or an accident, so the records of their lives were fragmented. There are no records in Camden from the point of view of the Chinese gardeners. The first mention of a Chinese death in Camden was in 1921. [36] Chun Yuen died in Camden Hospital after an accident when he encountered a mob of cattle being driven to market. Chun Yuen was taken to Camden Hospital by car, which had come along shortly after the accident. Dr. West attended the injured man at the hospital, but Chun Yuen died two hours later (*Camden News* 12 May 1921). The coronial inquest with A.E. Baldock found accidental death after Chun Yuen was 'knocked down by a cow from a mob of cattle on the Menangle Road, Campbelltown' after the cattle had been startled by dogs. His brother Chine Leung, identified Chun Yuen, who stated that Chun Yuen 'had no money here. He sends it along to China to his wife and family. He could not speak English and could not understand English'. (Witness statement, *Camden News* 12 May 1921).

Death Certificate

The Death Certificate (Figure 23) stated that Chun Yuen was a Chinese market gardener. The Certificate Registrar was Mr. Ben Hodge, who recorded that Chun Yuen was buried in the Chinese section of Rookwood Cemetery. The witnesses of the burial were P Peters and H.W. Crane. Chun Yuen was fifty-one years old and had been in the colony for thirty years. In 1888, he married Ung Lee in Canton when he was eighteen. He had come to the colony about 1891 when he was 21. [37]

On 16 March 1909, Chun Yuen, aged 54, applied for a Certificate of Exemption from the Dictation Test for up to three years. He travelled to China from Queensland on the ship *Taiyuan* and returned on the ship *Eastern* on 12 July 1910. [38] Travelling via Queensland was the cheapest way to return to China, but it would have been difficult. Chun Yuen is buried in the Chinese section of Rookwood Cemetery. He was buried on 2 May 1921. [39] His records combine the names of prominent Camden characters: Dr. West, registrar Ben Hodge, and witnesses P Peters and H.W. Crane.

Sadly, he came to Camden, even though he could not speak or understand English, hoping to make money for his family and died far from home. The immigration restrictions and demeaning handprints show that the Chinese endured discrimination, particularly in the Commonwealth government restrictions. Chun Yuen is remembered through the local paper and the National Archives, which tell the story of one of Camden's market gardeners.

---

[36] Births, Deaths and Marriages, YUEN (CHINESE) CHUN, 8821/1921.
[37] Chun Yuen Death Certificate, Registration Number 8821/1921.
[38] National Archives of Australia: Chun Yuen, J2483, 17/30.
[39] Rookwood, Chun Yuen, 1921 Allotment 266.

7 • Chun Yuen (1870-1921) – Accidental Death on the Road

Figure 22 - Chun Yuen's Certificate Exempting from Dictation Test, 1909 (National Archives of Australia J2483, 17/30. Page 1).

7 • Chun Yuen (1870-1921) – Accidental Death on the Road

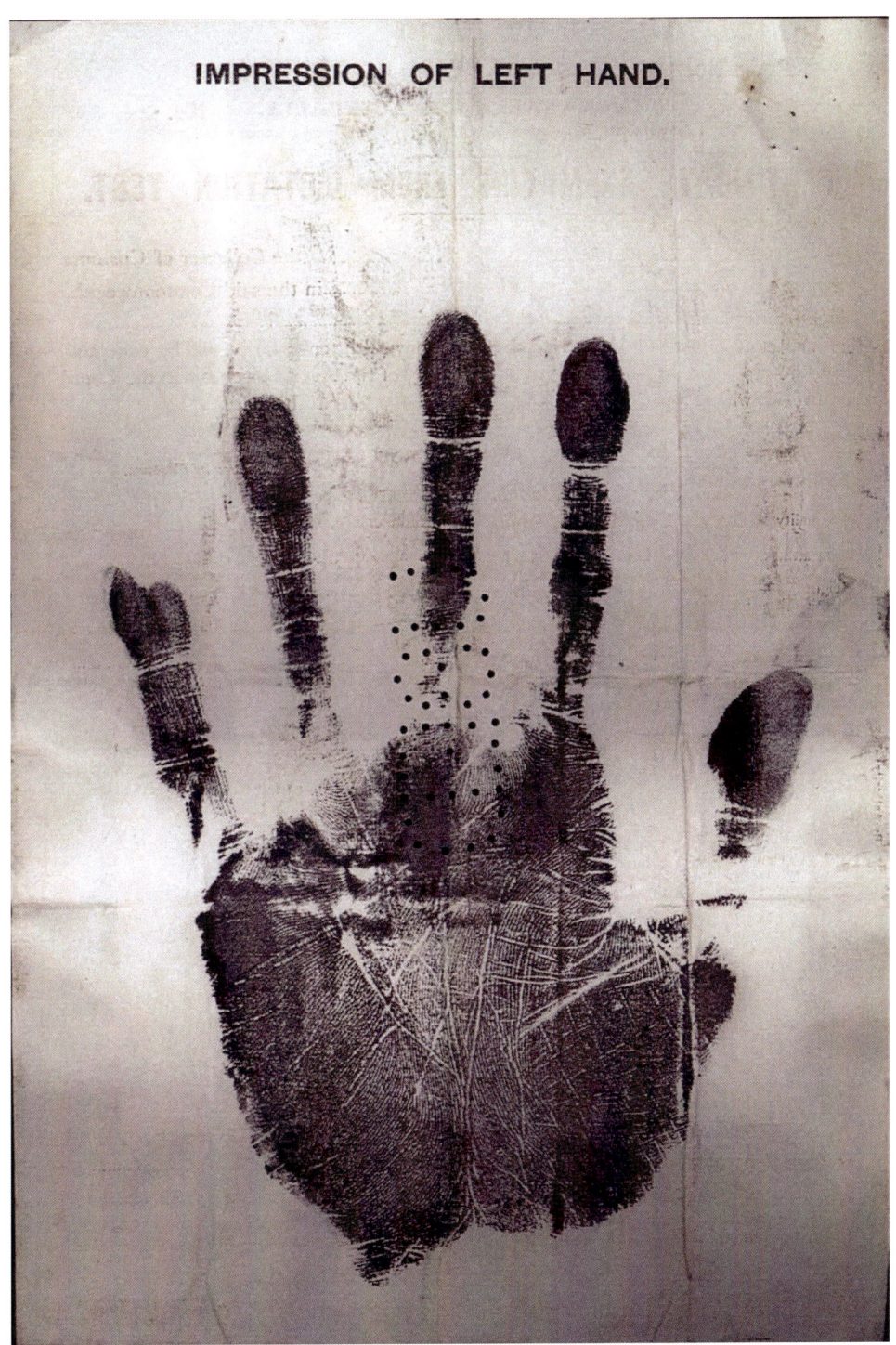

Figure 23 - Chun Yuen's left-hand print on his Certificate Exempting from Dictation Test, 1909. (National Archives of Australia: J2483, 17/30. Page 2).

NEW SOUTH WALES

BIRTHS, DEATHS AND MARRIAGES REGISTRATION ACT 1995

REGISTRATION NUMBER
8821/1921

DEATH CERTIFICATE

	Date and place of death	Name and occupation	Sex and age	Cause of Death Duration of last illness; medical attendant; when he last saw deceased	Name and occupation of father. Name and maiden surname of mother	Informant
271	1921. 30th April Camden District Hospital Camden Municipality N.S.W.	Chun Yuen Chinese Assistant Market Gardener	Male 51 Years	(1) Infra Cerebral Haemorrhage & Laceration of the Brain (2) Accidentally received (3) Finding at (4) Inquest held 4th May 1921 A.E. Baldock Coroner	Chun (1) Kum (2) Farmer (3) Kwok See	A.E. Baldock Coroner Camden

Particulars of Registration	When and where buried; name of undertaker	Name and religion of Minister and names of witnesses of burial	Where born and how long in the Australasian Colonies or States	Place of marriage, age, and to whom	Children of marriage
(1) R. Hodge (2) 4th May 1921. (3) CAMDEN	(1) 2nd May 1921 Chinese Section Rookwood Cemetery Rookwood (3) Walter F. Peters	(1) W. Todd (2) Church of England (3) F. Peters H. W. Crane	Cantong China N.S.W. 30 Years	(1) China (2) 18 (3) Ung See	years. Unknown

Before accepting copies, sight unaltered original. The original has a coloured background.

REGISTRY OF BIRTHS
DEATHS AND MARRIAGES

SYDNEY **29 February 2016**

I hereby certify that this is a true copy of particulars recorded in a Register in the State of New South Wales, in the Commonwealth of Australia

Registrar

Figure 24 - Death Certificate for Chun Yuen.

8 • Ah Chong

Killed in a Train Accident

Julie Wrigley

Ah Chong (c. 1860-1922) is a good example of the Chinese market gardeners in Camden in the early period before 1922. He was probably born in the Canton district of southern China about 1860 and came to Australia before the restrictions of 1901. Almost all Chinese market gardeners who came to Camden in the late nineteenth century were born in Guangdong Province and left China as young men. They saw themselves as Chinese working away from home until they made enough money to return.

From the report of his death in the local paper stating Ah Chong had lived in the area for fourteen years, it seems he came to the Camden district about 1908 (*Camden News* 1 June 2022). The first reference to Ah Chong is in the *Camden News* (30 September 1915), where Ah Chong donated one shilling to the farewell of volunteer soldiers attending the First World War. Occasionally, Ah Chong and other Chinese gardeners donated to Camden causes, particularly to the Camden District Hospital.

In the 1920s, all the gardeners' produce was taken by the Camden-Campbelltown train, Pansy, to the markets in Sydney. In the period before cars, the train was hugely important for transport. It brought manure to Camden from Sydney horse stables and took vegetables to the Sydney markets. The train allowed the Chinese market gardeners to come from Sydney via Campbelltown to Camden.

Unfortunately, Ah Chong was killed in a train accident in 1922. (*Camden News* 1 June 1922).

> Early on Monday morning last the body of a Chinese named Ah Chong was found on the railway line between the Camden Vale Milk Depot and the Camden railway bridge. The deceased, who was familiarly known as Willie, and whose age is given as sixty-two years, had lived in the district following the occupation for about fourteen years. It is quite evident from the condition of the body when found that he was struck and killed instantly by the Camden tram the previous evening. How he came to be on the line, though he had to cross it to reach the home of his Chinese friends, whom he frequently visited, remains a mystery. He was seen only a few minutes prior to the tram reaching Camden and it is generally surmised that he attempted to cross over the line ignorant of the approaching danger. The Camden police had the body removed to the hospital morgue, and it was later taken to Rookwood Cemetery.

Ah Chong was typical of the Chinese market gardeners in Camden in the early twentieth century. He was a male gardener living without close family but sending money home to his wife and children in China and saving for a trip back to China. His plan would have been to return to China to his ancestors' home at the end of his life.

8 • Ah Chong – Killed in a Train Accident

Figure 25 - Camden-Campbelltown train on Cowpastures Bridge, 1910 (Camden Museum).

The Coroner's Inquiry on 2 June 1922 revealed some details of Ah Chong's family (*Camden News* 8 June 1922). He had a brother, Ah Lung, living at 321 Church Street, Parramatta, and a cousin, Ah Sing, living at Camden. His cousin, Ah Sing, stated that Ah Chong had a wife and two children in Canton, China, but apart from his brother and cousin, no other relatives were living in Australia. The registration of Ah Chong's death states his father was Yen Duck, his mother was Wong See, and he was sixty-two when he died. [40]

Deceased Estate

Ah Chong's Deceased Estate record states that Ah Chong (also known as Sue Ah Chong) had £196 in the Government Savings Bank of NSW at Camden and £4 money in hand, making a total of £200 in his estate. The duty payable on that amount was £4. [41] The amount Ah Chong had in the bank would be more than he would have earned as a peasant farmer growing vegetables in China at that time.

Ah Chong died without a will, so the Public Trustee supervising the estate of Ah Chong placed a notice in the *Camden News* (22 June 1922) asking those who believed Ah Chong owed them money to lodge their claims within a month. The rest of the estate would then have been distributed.

---

[40] Death Certificate, NSW BDM, [5841/1922].
[41] NSW State Records, Kingswood, Deceased Estate Ah Chong, Public Trustee.

8 • Ah Chong – Killed in a Train Accident

Ah Chong was buried at Rookwood,[42] as many other Camden Chinese market gardeners were also. Ironically, the train made it easy for the market gardeners to travel to Camden for work, but it was also dangerous, and several Chinese men lost their lives in train accidents.

Figure 26 - Camden-Campbelltown train, Pansy, about 1917 at Camden Railway Station (Harold Perkins).

On Chong Sing

In 1939, another market gardener, On Chong Sing, was killed by a train. The *Camden News* (16 March 1939) recorded the accident.

Allen Wilfred Lowe, engine driver, stated that he was the driver on the 2.45 p.m. train from Camden to Campbelltown. As the train was approaching Elderslie at about 2.47 p.m. on the 25th of February, he sounded the engine whistle, and as there was no one on the Elderslie platform, he was not going to stop there, but just then, a Chinaman rushed out of the grass or from under a tree and tried to cross the line in front of the engine. An effort was made to stop the train, but as the man was only about five or six paces in front of the engine, it was impossible to stop in time to save him from being hit.

---

[42] Rookwood, Ah W Chong, date of burial 30 May 2022, Chinese 3_Zone H/#/236.

8 • Ah Chong – Killed in a Train Accident

Long Lee

In September 1943, Long Lee (also known as Joe Long), a Chinese gardener aged 68 working at the garden near the Camden Milk Depot, was killed instantly by being hit by a train. (*Camden News* 7 October 1943). The accident happened when Long Lee entered the railway crossing near the milk depot, and the train travelling to Camden struck him. The District Coroner, H. S. Kelloway, J.P., held an inquiry at the Camden Court House on the death of a Chinese man, Joe Long, who was killed on 21st September. The coroner's verdict read:

> The deceased died from internal injuries received on the 21st September when he was accidentally run down by a railway engine near Camden Railway Station. I further find that no blame could be attached to the staff of the train involved in the accident.

Summary

The Chinese market gardeners worked hard, hoping for success to benefit their families back in China, but unfortunately, some met their death by train accidents. The names of many of the Camden Chinese market gardeners are lost now, but some names were recorded in the local paper when there was an accident.

Figure 27 - Camden-Campbelltown locomotive Pansy and the loading siding near the former Milk Depot. Note how dangerously close the train lines are to pedestrians (Camden Museum).

9 • Lowe Kum Ming and Ah On

Tragedy of Murder and Suicide

Julie Wrigley

The Chinese market gardeners led peaceful lives in Camden for over ninety years, and in all that time, there was only one serious crime. The newspapers reported the tragic event in 1935. One headline said, 'Terrified Township. Murder at Camden. Chinese Killer Causes Panic' (*Daily Mercury*, Mackay Qld, 27 May 1935), though that was not a local paper. The event concerned a murder and a suicide and has several sad aspects.

The Chinese man who committed the murder, Ah On, was born in Canton in 1871 and came to the colony in 1892, before the immigration restrictions of 1901. According to immigration records, Ah On worked as a market gardener at Gosford for six years, Smithfield for four years, and Fairfield for about thirteen years. [43] He had gone back to China for two visits. One trip was from February to December 1919, and the second was from June 1922 to January 1924. In 1926, he was given a reference stating that he had worked for Sun Chong Kee garden at Camden for the past fifteen months. He was known in Camden as 'Smiler', but he was undoubtedly lonely as his wife and three children were back in Sun Jing, China, and he was responsible for sending them money.

Mental Illness

A few weeks before the murder, Ah On became sick and confused. At the inquest into the murder, his cousin said that Ah On had been daydreaming of 'growing wings' and 'flying back to China' as he did not have enough money for the trip. Ah On had been drinking heavily and was sure he had

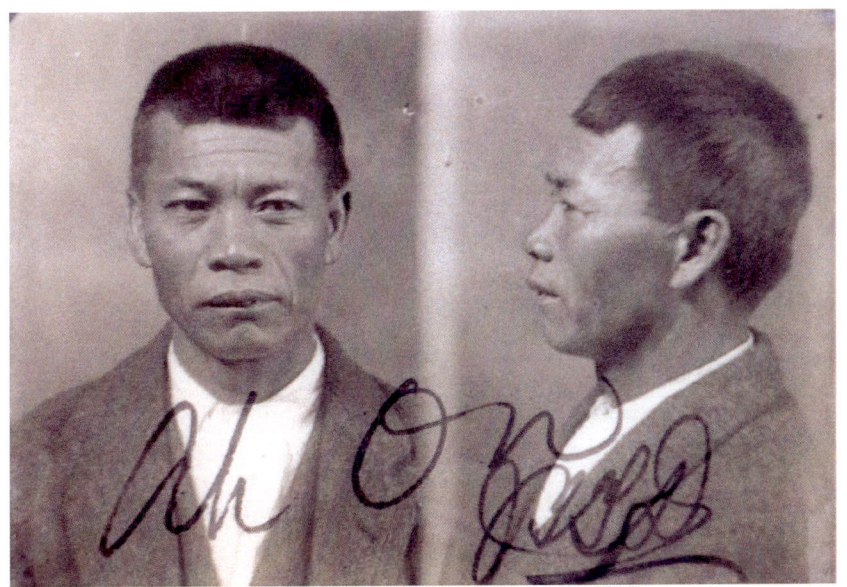

Figure 28 - Ah On, Immigration papers, 1922. (National Archives of Australia: Certificate Exempting from Dictation Test ST84/1, 1922/333/91-100. Page 7).

---

[43] National Archives at Chester Hill, Ah On.

9 • Lowe Kum Ming and Ah On - Tragedy of Murder and Suicide

an enemy, though this was a mistaken belief. Ah On meant to punish Sun Chong, but he struck the wrong man. (*Picton Post* 13 June 1935).

The newspapers revealed the details. The event happened about nightfall on Friday, 24 May 1935, at the garden near the Camden Weir about a mile east of the Macquarie Grove Road, at Sheil's garden. The group of ten gardeners were leaving the packing shed where they had been loading vegetables for the city market onto the truck of the carrier Herbert Moss and then planning to have their evening meal at their nearby hut. At dusk, they were coming along a path between vegetable beds, and Low Kum Ming was the head of the line. Lowe Kum Ming was struck by a chopper and killed, but the assailant fled. (*Picton Post* 13 June 1935)

The Chinese gardeners ran for help to the Sheil's house, *Caernarvon*, and he rang the Camden police, Sergeant E.H. Porteus and Constable Harry Haylock, who were well-liked in the town. Other police came from Sydney and an official interpreter because some Chinese gardeners did not speak English. The gardeners realised Ah On was missing, but no one knew where he had gone. There was a huge search for ten days. Newspapers nationwide had a series of sensational headlines during the search: 'Chinese Murdered in Camden Market Gardens. Compatriots' Amazing Story' (*Sydney Morning Herald* 25 May 1935); 'Chinese Hacked to Death. Camden Chopper Murder. Four Terrific Blows. Eight Terrified Gardeners' (*The Daily Telegraph*, 25 May 1935); 'Chopper Crime. Chinese Gardener Hacked to Death. Thirty Detectives Search for Assailant and Weapon. Camden' (*Daily Advertiser* Wagga Wagga 27 May 1935); and 'Camden Murder. Missing Chinese Found in River' (*Canberra Times* 4 June 1935.)

The body was found on 3 June, but it was estimated that the drowning had occurred nine or ten days earlier, immediately after the death of Lowe Kum Ming. The newspaper *Truth* (9 June 1935) explored the story in detail using colourful language 'Camden Killer's Obsession – Strange Story at Inquest into Deaths of Two Men'. *The Picton Post* (13 June 1935) had the story of 'perhaps the most sensational crime in this district's history' with the heading 'Coroner's Verdict – in Chinese Chopper case'.

Lowe Kum Ming (or Low Kum Ming) was fifty-five when he died. According to his gravestone, his family name was Liu (Lau), and his given name was Jin Ming (Kum Ming). [44] He was born in Zhongshan about 1880 and was issued an Alien Registration Certificate in Sydney in 1919. [45] He had gained a Certificate Exempting from Dictation Test in 1932. [46]

The sad part for Lowe Kum Ming's family is that he was innocent of wishing any harm to Ah On. Lowe Kum Ming was a poor man, aged 55, married to a wife and three children in China.

Deceased estate papers at NSW State Records, Kingswood, show he had £3 in cash, £22 in unpaid wages and £10 as a refund of his share in the Grove market garden, making an estate of £35. [47] Lowe Kum Ming did not have a will. The *Camden News* (12 September 1935) published a

---

[44] Remembering the Forgotten: Chinese Gravestones in Rookwood Cemetery 1917-1949 Jones, Doris Yau-Chong.
[45] *National Archives of Australia:* Certificate No. 1886, BP4/3 Chinese – Ming Kum.
[46] *National Archives of Australia:* Kum Ming, SP42/1, C1932/2547.
[47] *State Records NSW*, Kingswood: NRS 13340, Deceased estate files, Lowe Kum Ming, date of death 24 May 1935, Pre A 94764 [20/2016] and NRS 13340, Deceased estate files, Garn On, date of death 24 May 1935, Pre A 110439 [20/2208].

9 • Lowe Kum Ming and Ah On - Tragedy of Murder and Suicide

statement from the Public Trustee that those having claims on the estate needed to lodge them within two months. The assets would then be distributed to those entitled to benefit.

Another sad part is that the murderer, Ah On, committed suicide by drowning himself in the river near the weir. *The Sydney Morning Herald* (4 June 1935) reported that his body was found by two schoolboys, Allen Boardman and Edgar Johnson, who were canoeing on the river near Macquarie Grove Bridge and who reported the find to the police.

Figure 29 - Innocent victim Kum Ming, 1932. (National Archives of Australia, NSW Office Chester Hill SP42/1, C1932/2547.

9 • Lowe Kum Ming and Ah On - Tragedy of Murder and Suicide

Inquest

At the inquest conducted by the coroner, [48] A.E. Baldock, at the Camden Courthouse, the Government Medical Officer, Dr. Crookston, described the terrible injuries inflicted on Lowe Kum Ming. Arthur Chong, the principal shareholder and manager of the garden, stated that Ah On had one full share, valued at £51 in the Macquarie Grove garden, less £15 for money advanced to him. The coroner praised the police for the support and protection they had given the Chinese gardeners when they feared the attacker. In 1935, there was also a magisterial inquiry, [49] and in 2019, an interview with Christopher Evans, retired Assistant Police Commissioner, and Ian McCrae, NSW Magistrate, featured in the DVD film about the Chinese market gardeners in Camden made by Wen Denaro, *100 Yards of Silk*. [50]

An interesting article from *The Daily Telegraph* (6 June 1935) had the heading 'Lonely Funeral – Chinese Murderer Shunned',

> The bodies of Low Kum Ming [or Ko Ming], the murdered man, and Ah On [or Karg On], the suicided murderer, lie side by side, in Rockwood Cemetery. These two – the one whose head was split open by many blows of a chopper at Camden last week, and the man whom they say wielded the chopper. In death they are neighbours. But it was sunny when they buried Kum Ming with honour. It was cold and bleak and lonely when the body of Ah On was lowered into its grave.
>
> Kum Ming had many to wish his soul godspeed on the journey to join his celestial fathers. He was a poor man, and so there was no roast pig, he had many little comforts…He had a good send-off with fine red candles and joss sticks at the foot of his grave, some sweetmeats, toffee and nuts, and some words reverently spoken before the coffin was lowered. They burned wafers and there was one man there who took out a handful of threepenny bits and handed them round to all the others and said 'Take these for luck'… Also Ko Ming's brother wept.
>
> But there were only Mr. Butler, the undertaker from Camden, and his assistant and a gravedigger when Ah On was buried. When it was finished, at the foot of Low Kum Ming's mound was a board properly inscribed with characters but at the foot of Ah On's mound there was none. By his honourable suicide, which removed the stain of blood from his hands, the soul of Ah On was redeemed – as far as his ancestors were concerned. He is even now with those ancestors…

Their bodies lie so close together, but Kum Ming's ancestors and Ah On's ancestors may abide in very different parts of the infinite celestial garden.

Rookwood Lowe Kum Ming is in grave 746, and Ah On is in grave 747. [51] Lowe Kum Ming has a

---

[48] *State Records NSW*: NRS 343, Coroner's inquest papers, 1935 [2/10529] File No.872 of 1935 re Ming Lowe Kum and Garn On.
[49] Magisterial inquiry, 1935 [3/959, entry No.872.]
[50] DVD *100 Yards of Silk*, 2016, directed by Wen Denaro and Bernie Zelvis. Camden Council gave a grant of $5,000 to the Camden Historical Society towards the cost of making the DVD.
[51] Chinese Section 3 Zone H.

9 • Lowe Kum Ming and Ah On - Tragedy of Murder and Suicide

headstone, but Ah On does not. The story of these two men is a contrast, but it shows the mental strain of loneliness and isolation that the Chinese market gardeners endured far from their homes. The tragedy was outside the normal, peaceful lives of the Chinese market gardeners.

Figure 30 - Rookwood, Lowe Kum Ming headstone lower right corner (John Wrigley, 2016).

10 • Wong Yong and Trips Home

Julie Wrigley

Many Chinese gardeners must have longed for home and dreamed of winning the lottery to enable them to go home to their families. In 1946, two lucky Chinese market gardeners in Camden did win first prize in the State Lottery. In June, Wong Yong won £5000; by extraordinary coincidence, Willie Chung won £5000 in November.

The *Camden News* (27 June 1946) stated that Wong Yong had won first prize in the State Lottery and went on to say,

> Wong Yong, known as "Stumpy", had donated £50 to the Camden District Hospital and £50 to the Red Cross. He had left the district prior to the drawing of the lottery in preparation for his return to China. He was at present waiting for a ship sailing to Hong Kong.

Lottery Winning Celebration

The Sun (4 July 1946) explained,

> The 60-year-old Wong Yong held a celebratory meal for 150 friends consisting of a ten course dinner at a Campbell Street Chinese café. It cost him £300. The guest of honour was Archie Tippetts, a Camden carrier, who had bought the ticket in his own name because Wong Yong could not read or write English.

Figure 31 - Archie Tippetts' truck in Elizabeth Street, laden with cauliflowers, grown by local Chinese, 1932. One truck has the sign "AC Tippetts Camden – Sydney" (Camden Museum).

10 • Wong Yong and Trips Home

The granddaughter of Wong Yong, Daphne Lowe Kelley, is well-established in Sydney's Chinese community and has played a leadership role in many Chinese Australian organisations. She is a founding member and former president of the Chinese Heritage Association of Australia and the current Chair of the Museum of Chinese in Australia's Board of Directors. She visited her maternal grandfather's village, Bi Toa/Pitao, in 2018.

She recently supplied details about her maternal grandfather, who was born on 15 November 1879.

His full name was Chung Wong Ying (Chung is his surname), and he came from Bi Toa/Pitao Village in Jung Seng /Zengcheng County, Guangdong Province. Zengcheng has now been incorporated as a District in Guangdong's capital city of Guangzhou.

Wong Yong came to Australia in the late nineteenth century before Federation and the *Immigration Restriction Act 1901*. While in Australia, he always worked as a market gardener, and the last garden he worked on was the Hop Chong Garden, later owned by Biu Wong, who was from the same county.

In 1946, Wong Yong returned to China with his family but died there in February 1947. He left some of his winnings in a Camden bank account. Daphne Lowe Kelley said that many years later, the family was able to retrieve the money, but unfortunately, as it was a cheque account, fees had been deducted, and there was no interest.

Wong Yong's name is on the Honour Board at the Camden District Hospital (Figure 32). The board acknowledges the generosity of Wong Yong and Willie Chung (see chapter 11) but, unfortunately, has misspelt Wong Yong (recorded as Wong Tong) and Willie Chung (recorded as Chong). It seems the authorities did not take much care with the Chinese names.

Figure 32 - Chung Wong Ying also known as Wong Yong (Daphne Lowe Kelley).

11 • Willie Chung

Australian-Born Chinese Leader

Julie Wrigley

The Camden District Hospital Life Members Honour Board is still at Camden Hospital. It records the names of two of Camden's Chinese market gardeners, Wong Yong and Willie Chung (Chong), as a mark of respect for their generosity to the Hospital after the men separately both won the State Lottery in 1946.

Figure 33 - Camden District Hospital Life Members Wong Yong (Tong) and Willie Chung (Chong) (John Wrigley, 2017).

The life of Willie Chung (1897-1966) is well documented in the National Archives of Australia (SP42/1 C1946/5927), with ninety-seven pages in the file. He was born on 19 September 1897 at Cross Creek, Nemingha, near Tamworth, New South Wales. His elder brother Charlie was born in 1896. Their father was Charlie Chung, a Chinese tobacco grower and gardener of Nemingha, Tamworth. Their mother was Mary Hilda Brown. Willie's birth certificate [52] says his father was 32, and his mother was a 19-year-old 'spinster' when he was born, so the boys were born out of wedlock. A note on Willie's birth certificate explains that on 16 August 1899, Charlie and Hilda were married. However, a formal Deed of Separation was drawn up on 27 October 1899 [53].

In 1899, when Charlie was three years old and Willie was two years old, their father, Charlie Chung, sent his sons with his uncle to China. They travelled on the ship *Airlie*. Charlie gave a

---

[52] *National Archives of Australia*. Willie Chung SP42/1 C1946/5927, Birth Certificate 89, 90, 91/97.
[53] *National Archives of Australia*: Willie Chung SP42/1 C1946/5927, pages 89, 90/97.

11 • Willie Chung – Australian-Born Chinese Leader

Statutory Declaration in 1912 that he sent his sons 'to be educated in the Chinese language and customs by his father and mother'. [54]

Red Tape and Racism on Return

Willie (15½ years old) and Charlie (16½) arrived back in Sydney on 23 November 1912 on the *SS Changsha*. The re-admission of Willie and his elder brother Charlie to Australia in 1912 caused a great deal of red tape between Customs and the Department of External Affairs, who were concerned with finding out if the boys were 'half-cast' or 'full Chinese' and whether they were legal immigrants. There were urgent telegrams sent between the Boarding Inspector and the Secretary to the Department of External Affairs. The authorities organised a doctor to examine the boys to determine whether they were 'half-caste' as they claimed and whether they were the same boys who had left in 1912 (Figure 34).

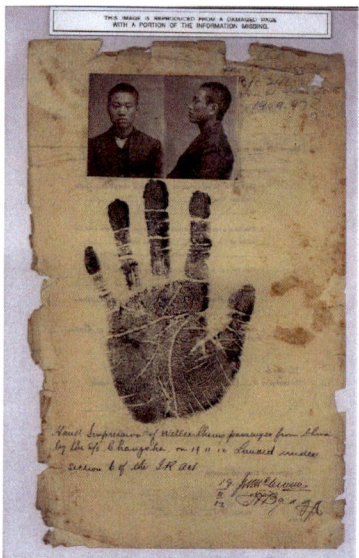

Figure 34 - Willie Chung, Immigration papers, 1912
(National Archives of Australia SP42/1 C1946/5927, page 85/97).

The boys were finally admitted to Australia under section six of the Immigration Restriction Act. Charlie Chung paid a deposit of £200 for entry for one month, which would be refunded if the boys were who they claimed to be. The father brought three persons to Sydney to confirm the identity of his sons: Mrs. Jane Woods, a neighbour; Mrs. Annie Long, the European wife of another Chinese man; and Mr. Alexander Matheson, a former mayor of Tamworth. They all made statutory declarations about the boys' identity, and the boys were readmitted to Australia after much discrimination and racism. [55]

---

[54] *National Archives of Australia*: Willie Chung SP42/1 C1946/5927, 75/97 pages.
[55] *National Archives of Australia:* Willie Chung SP42/1 C1946/5927, 65-69/97 pages.

11 • Willie Chung – Australian-Born Chinese Leader

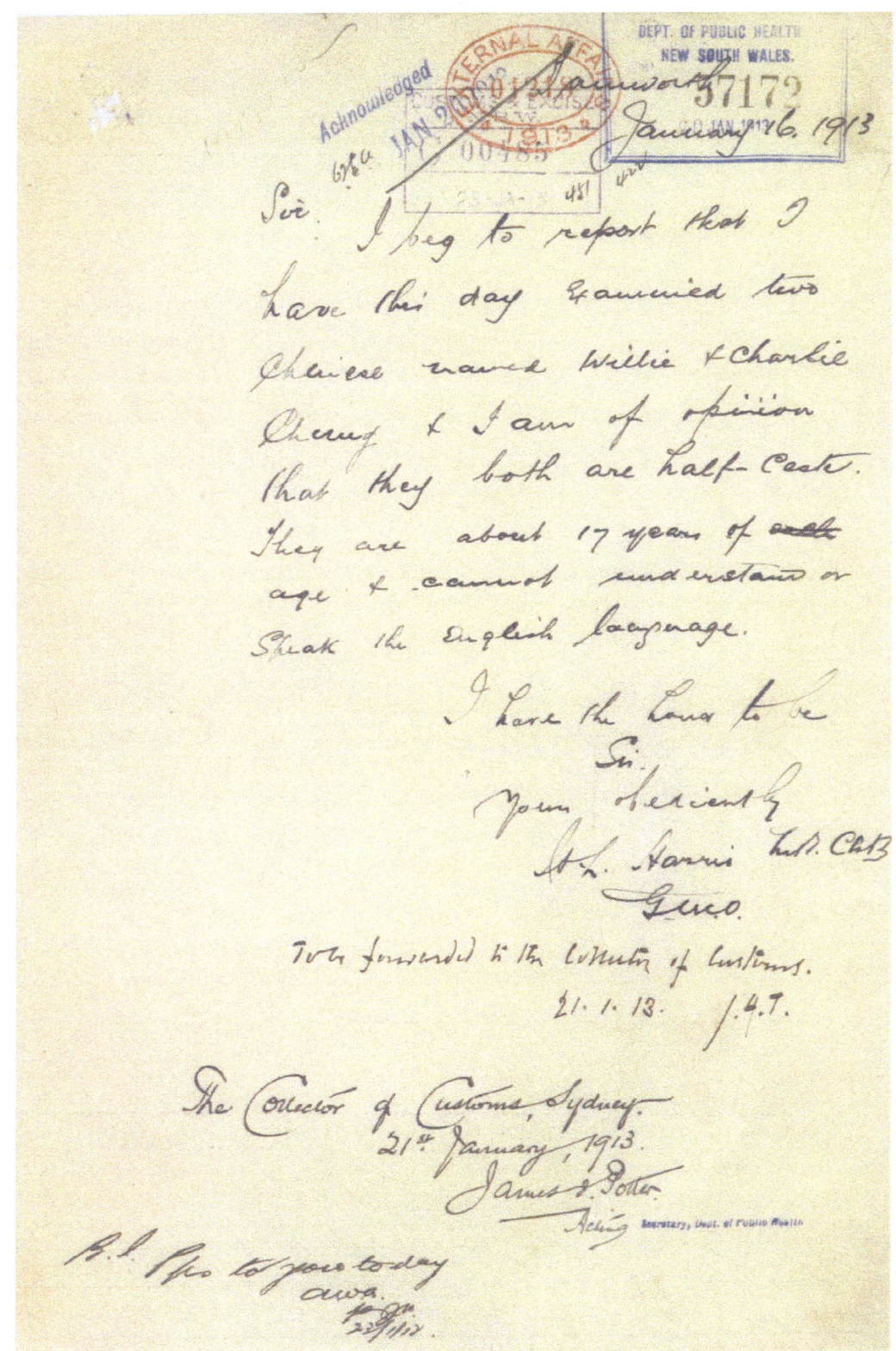

Figure 35 - Doctor's letter confirming Willie and Charlie Chung were 'half caste' (National Archives of Australia: SP42/1 / C1946/5927, page 50/97).

11 • Willie Chung – Australian-Born Chinese Leader

When the boys arrived in Sydney, they barely spoke English. Willie worked on his father's tobacco farm at Tamworth for four years before he returned to China in 1916 for another three years there. During those years, Willie married Toong Goon in Canton, and she bore him three sons: Leon/Leong Poy, Hom Poy, and Ah Lum. In 1922, the Tamworth Solicitors, Tribe and Prentice, stated about Willie's father,

> Charlie Chung has been a client of ours for very many years, and is one of the most highly respected Chinese farmers in this district, and bears an excellent character in every respect. [56]

It seems Willie Chung left Tamworth, possibly after his father died in 1924 and worked at gardens in Sydney. In 1931, Willie visited China and returned the same year. Willie had a daughter, Ah Hoe, in 1931. He made a trip to China in 1937 for six months, and there are references from St. Ives, Sydney, stating he had worked there for five years. Willie came to Camden and joined the Chinese market gardeners in the 1930s. Richard Nixon referred to Willie Chung at the Camden Historical Society meeting on 9 October 1996 (Museum archives):

> He was an honourable and well-respected Chinaman, in charge of Sun Chon Key's garden for about thirty years. He suffered a severe infection on the back of his neck, probably a carbuncle. In common with tradition, the local Chinese community did not have much faith in Western medicine, preferring their own traditional herbal medications. Willie Chung was sure he would die, for had not his cousin of the Sydney markets had the same complaint and, having been taken to Sydney Hospital – and operated upon – died within three days? Willie was sure that if he was operated upon, he would die in three days! One Sunday morning, his associates placed a mattress on the bottom of a spring cart to not hurt Willie and very slowly walked the horse into town to Doctor F.W. West. There they told their story – Willie would die in three days if operated upon. The doctor assured them there would be no operation if Willie went to the hospital – so he went to the hospital for a long while. Fifty years ago, there were no antibiotics, but Willie recovered by skilful and careful treatment. His way of saying 'thank you' was a substantial donation (for those days) to the hospital – hence his name on the Honour Board.

Willie Chung was listed in the 1943 Electoral Roll as working at Yee Lee vegetable garden, but in 1946, at the time of his lottery win, he was listed at Hop Chong Garden. In April 1946, Willie applied for a Certificate of Exemption to allow him to visit China for thirty-six months and be readmitted to Australia. [57] He made a Statutory Declaration that he lived in Tamworth, Camden, and Sydney and currently resided in Dixon Street, Sydney. He stated that he was married to Toong Goon, Canton, China, and had been to China five times. He had left Australia in 1899 and returned in 1912; left in 1916 and 1919; left in 1922 and 1923; left in 1931 and returned in 1931. On 13 January 1937, he left on the *SS Tanda* and returned on 21/7/1937 on the *Nankin*. [58]

---

[56] *National Archives of Australia:* Willie Chung SP42/1 C1946/5927, page 28/97.
[57] *National Archives of Australia:* SP42/1. C1946/5927, Willie Chung, 4/97 pages.
[58] *National Archives of Australia*: Willie Chung, SP42/1 C1946/5927, page 50/97.

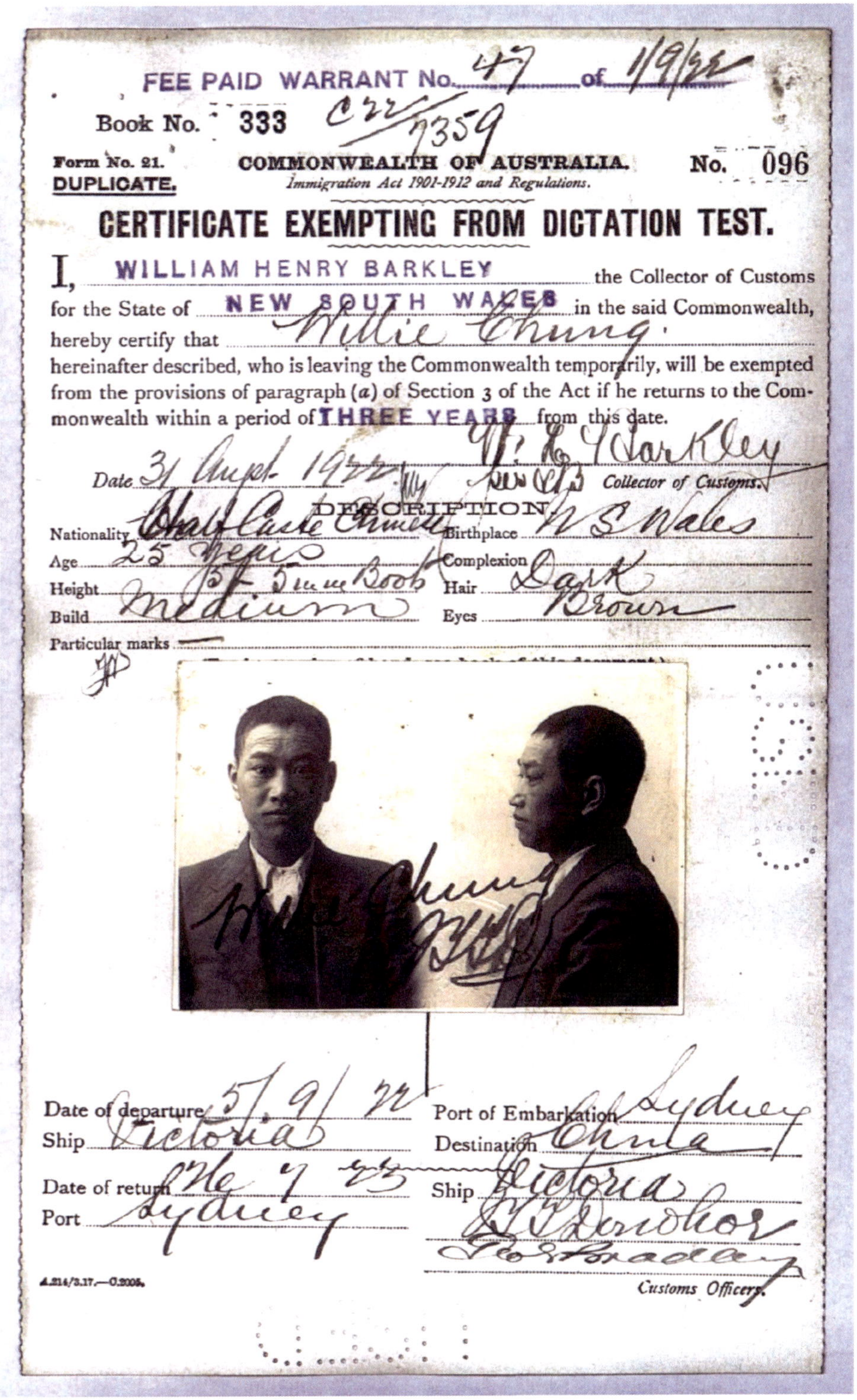
FEE PAID WARRANT No. 47 of 1/9/22

Book No. 333 C22/7359

Form No. 21. COMMONWEALTH OF AUSTRALIA. No. 096

DUPLICATE. *Immigration Act 1901-1912 and Regulations.*

CERTIFICATE EXEMPTING FROM DICTATION TEST.

I, WILLIAM HENRY BARKLEY the Collector of Customs for the State of NEW SOUTH WALES in the said Commonwealth, hereby certify that Willie Chung hereinafter described, who is leaving the Commonwealth temporarily, will be exempted from the provisions of paragraph (*a*) of Section 3 of the Act if he returns to the Commonwealth within a period of THREE YEARS from this date.

Date 31 Augt 1922 W. H. Barkley *Collector of Customs.*

DESCRIPTION

Nationality Half Caste Chinese Birthplace N S Wales

Age 25 years Complexion Dark

Height 5ft 5in in Boots Hair Dark

Build Medium Eyes Brown

Particular marks —

Date of departure 5/9/22 Port of Embarkation Sydney

Ship Victoria Destination China

Date of return 2/12/23 Ship Victoria

Port Sydney

Customs Officers.

A.214/3.17.—C.2005.

Figure 36 - Willie Chung's Certificate Exempting from Dictation Test, 1922 (National Archives of Australia: Willie Chung ST84/1, 1922/333/91-100).

11 • Willie Chung – Australian-Born Chinese Leader

Willie Chung Letter of Recommendation from the Mayor, 1946

His application was supported by a typed letter on 5 January 1946 from the mayor of Camden, H.S. (Stan) Kelloway, stating:

> To whom it may concern:
>
> This is to certify that I have known the bearer of this reference, Willie Chung, for a period of about eight years. During that whole of this time I have been very closely associated with him in both business and private life and have seen him almost daily. He bears an excellent character, is energetic, industrious, strictly honest and his integrity is beyond question. I might add that he is an abstainer. The respect in which he is held by his countrymen is fully shared by his European friends and acquaintances. I know him to be worthy of every confidence.

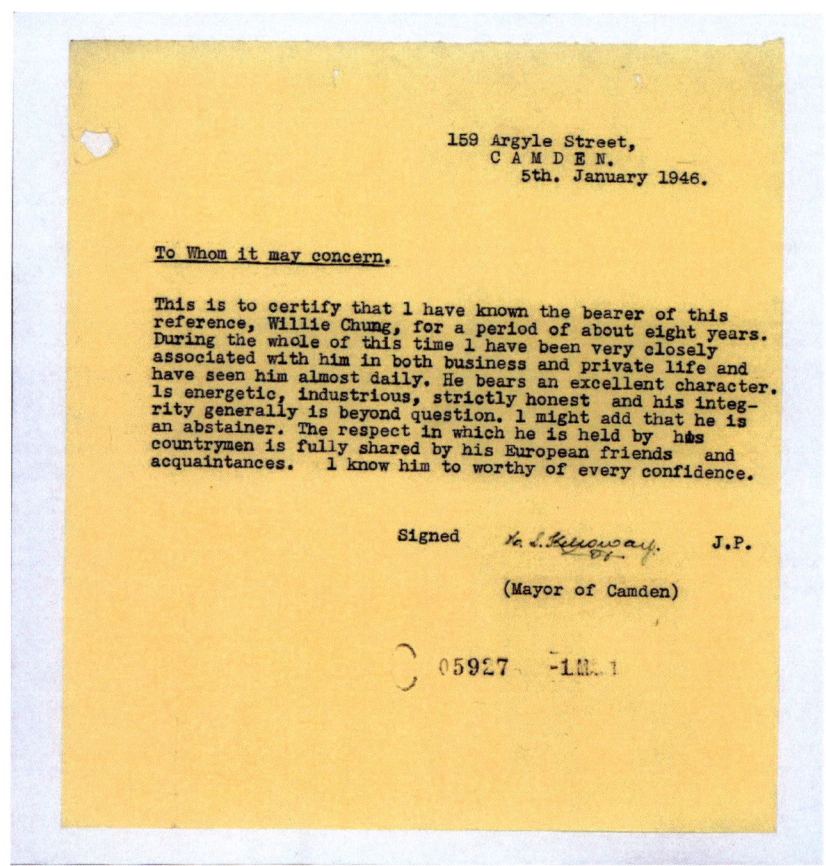

Figure 37 - Mayor Kelloway's letter of support (NAA: SP42/1, C1946/5927 page 5/97).

Another letter of support was from Bert Price, 'Riverview', Chellaston Street, Camden, a district agent for rotary hoes. [59]

---

[59] NAA: SP42/1, C1946/5927. Page 6/97.

In 1946, the children of Willie Chung were listed as Leong or Loon Poy (son b. June 1917), aged 28; Hom Poy (son b. July 1918), aged 25; Ah Lum (son. b. April 1919), 23 years; and Ah Hoe (daughter) aged 14. [60]

Lottery Winner

In November 1946, Willie Chung and his syndicate won a large first prize. (*Camden News* 28 November 1946).

> Camden agency for State Lottery tickets reports having sold ticket 57876 which won first prize of £5000 in the 1421 drawing yesterday. The winner is a 64 [in fact 49] year old Chinaman, W. Chung, of Hop Chong gardens, Camden. [The ticket] was named 'Black Cat'. This is the third success enjoyed by the same Chinaman during the past three months. He previously won £30 and £10 prizes from tickets ordered at the local lottery agency.

The money would have been distributed amongst the winning syndicate. Undoubtedly, some money went to Willy's wife and family in China, and perhaps some went into gambling.

In 1947, Willie Chung (Chong) and Wong Tong donated £50 each to Camden District Hospital, and their names are on the Hospital Honour Board for Life Members. Richard Nixon has stated in the Museum archives:

> In the closing years of his life, [Willie] brought his wife to live with him at the garden. This was not favourably received by some of his group, but such was their respect for him that it was tolerated.

Willie Chung died in March 1966 at Gladesville and was buried at Rookwood. [61]

Figure 38 - Grave of Willie Chung at Rookwood (John Wrigley, 2016).

---

60 National Archives of Australia: Willie Chung, SP42/1 C1946/5927, page 11/97.
61 Rookwood 4_Zone H/#/525.

11 • Willie Chung – Australian-Born Chinese Leader

Figure 39 - Willie Chung, 1937 (National Archives of Australia).

Willie Chung is a good example of a successful life lived by an Australian-born Chinese market gardener in Camden, where he was well-respected. Willie was treated with appalling discrimination and racism at the government level. Still, the Chinese and non-Chinese lived alongside each other in Camden with goodwill from the residents who appreciated the fresh vegetables the market gardeners provided. Many old-time Camden residents spoke warmly about the hard-working Chinese gardeners and memories of their surviving floods at Camden.

Apart from the name on the Camden District Hospital Honour Board, the name of Willie Chung is not remembered, but there are two names of Chinese market gardeners whose names are remembered by residents: Hop Chong and Biu Wong. Their stories follow.

12 • Hop Chong

The Best-Known Chinese Name in Camden

Julie Wrigley

Hop Chong, the Person

Hop Chong was one of Camden's Chinese market gardeners who worked along the river flats from 1910 to 1940. According to his gravestone, Hop Chong was born in Gao Yao, one of the districts in the province of Guangdong. [62] His family name was Zhu (Chu), and his given name was Yi Shen (Yee Sum). According to his death records, his father's name was Hay Wing, and his mother's was See Chang. [63] In Camden, he was one of Thurn's Chinamen working in Elderslie in the garden under the present Macarthur Bridge. Hop Chong applied for a licence for a pump on the Nepean River in 1914. (*Government Gazette* 14 January 1914).

Newspaper records show that Hop Chong, as an individual, contributed generously to the Camden community. He made several donations of money and vegetables. In September 1914, some Chinese market gardeners, including Hop Chong, donated four guineas each to the Camden Patriotic Fund (*Camden News* 10 September 1914). Hop Chong was a regular contributor to the Camden Cottage Hospital Appeals. He donated lettuce to the Hospital Appeal (*Camden News* 7 October 1915) and donated vegetables to Camden District Hospital Appeal (*Camden News* 11 April 1918). He donated two guineas to the Hospital Appeal (*Camden News* 11 July 1918), again in 1920 (*Camden News* 1 July 1920), and in 1921 (*Camden News* 23 June 1921). In 1928 (*Camden News* 13 December 1928), he donated five shillings to the Australian Cricketers Jolly Good Sports Roll for the Hospital Appeal. The following year (*Camden News* 31 January 1929), Hop Chong again donated two guineas to the Hospital's Annual Appeal.

Possibly about 1930, Hop Chong moved from the Elderslie garden (Thurn's) under the present Macarthur Bridge to the garden near the Cowpastures Bridge, closer to the town. *Camden News* (3 April 1930) reported that two men, Robert Hallam and Norman Gates, had been convicted of robbing the Chinese dwelling of Mee King, who was employed at Hop Chong Gardens at Elderslie. The men were identified, and each was fined £10 in default four months of hard labour. Interestingly, in the account of the robbery, Mee King said that he left the house at 5 a.m. to start work. The gardeners worked long hours with great commitment.

Hop Chong died in 1940 and was buried at Rookwood. [64] His headstone is recorded in the book by Doris Yau-Chong, *Remembering the Forgotten: Chinese gravestones in Rookwood Cemetery 1917-1949*. Unfortunately, many of the details on the headstones she recorded have weathered and can no longer be read.

---

[62] Remembering the Forgotten: Chinese gravestones in Rookwood Cemetery 1917-1949, Jones, Doris Yau-Chong, Invenet Publishing Pymble, 2003.
[63] Births Deaths and Marriages. Deaths: No.13113 /1940.
[64] Rookwood, Hop Chong, Zone H Chinese 3 Zone H/#/941. Buried 13 July 1940.

Hop Chong Garden as a Company

Hop Chong was used as a company name that lasted in Camden for decades. Many residents bought vegetables from the Hop Chong Garden Shop on the river flat north of Camden Valley Way and west of Cowpastures Bridge.

The Chinese gardeners coped with floods and fires. (*Camden News* 16 August 1951)

> On Thursday evening about 6 o'clock a fire broke out in a shack occupied by Hop Chong market gardens, Elderslie, and completely destroyed the building and contents. Six members of the company occupied the dwelling and were having their evening meal when a bright light was noticed in the kitchen. Willie Foote, one of the occupants, saw smoke coming from a corner of the building. He threw water where he saw the smoke, but suddenly flames shot up, which were fanned by a strong westerly wind, the whole place being gutted within a matter of minutes. Inquiries made indicate an electrical defect was the cause of the outbreak, there being no suspicious circumstances. The police got in touch with the Welfare Department which supplied food, clothing and blankets to the unfortunate Chinese.

The incident shows how many difficulties the Chinese market gardeners faced. Camden residents were sympathetic because people could see how hard the gardeners worked. In 1951 the group of seven Sun Hop Chong Company members (Willie Fook, Willie Chung, Ah Chang, Ah Kong, Chong Lung, Ernie Fatt and Lee Hing) raised £22 for the Camden Hospital Equipment Fund. (*Camden News* 25 January 1951). The popular Sergeant Dawson collected from all garden cooperatives and raised £160. Camden authorities and the Chinese gardeners cooperated, working for a good cause: the hospital.

In 1957, the Hop Chong Garden area was purchased from Whitemans by five market gardeners: Yung Sum Wong, Ho Jo Ling, Leung Ah Sin, Lo Lin Chow and Wong Tim (but Leung Ah Sin died in 1958). [65] In 1957, the Hop Chong group applied for a pump on the Nepean River to irrigate twenty-three acres. [66] In the 1960s, a number of the Chinese from the Hop Chong group gained Certificates of Naturalisation. The Commonwealth of Australia Gazette records Certificates were granted to Ho Jo Ling on 9 November 1961, Lo Lin Chow on 26 July 1962, Kwan Yin Tai on 6 May 1965, and Yuen Chi Sing on 17 June 1965. [67] Things were changing over time compared to the earlier government restrictions.

Mr. Cheuk Biu Wong purchased the Hop Chong Garden Company in 1968, with the financial help of his father (Yung Sum Wong) and gardened there with his family until 1993. The name Hop Chong Garden lived on in Camden long after the other Chinese market gardeners had gone.

---

[65] Births, Marriages and Deaths NSW, Leung Ah Sin died in 1958. Registration 26835/1958.
[66] Government Gazette Notices, 24 May 1957.
[67] Government Gazette Notices, 9 November 1961, 26 July 1962, 6 May 1965, 17 June 1965.

12 • Hop Chong – The Best-Known Chinese Name in Camden

Figure 40 - Yung Sum Wong and Siu Wong (the parents of Biu Wong) with their granddaughters Anne and Dianne and Sim and Biu Wong at Hop Chong Garden, 1969 (Photo supplied by the Wong family to Camden Museum).

Figure 41 - Sim Wong and baby Anne at Hop Chong Garden in 1967 (Photo supplied by the Wong family to Camden Museum).

12 • Hop Chong – The Best-Known Chinese Name in Camden

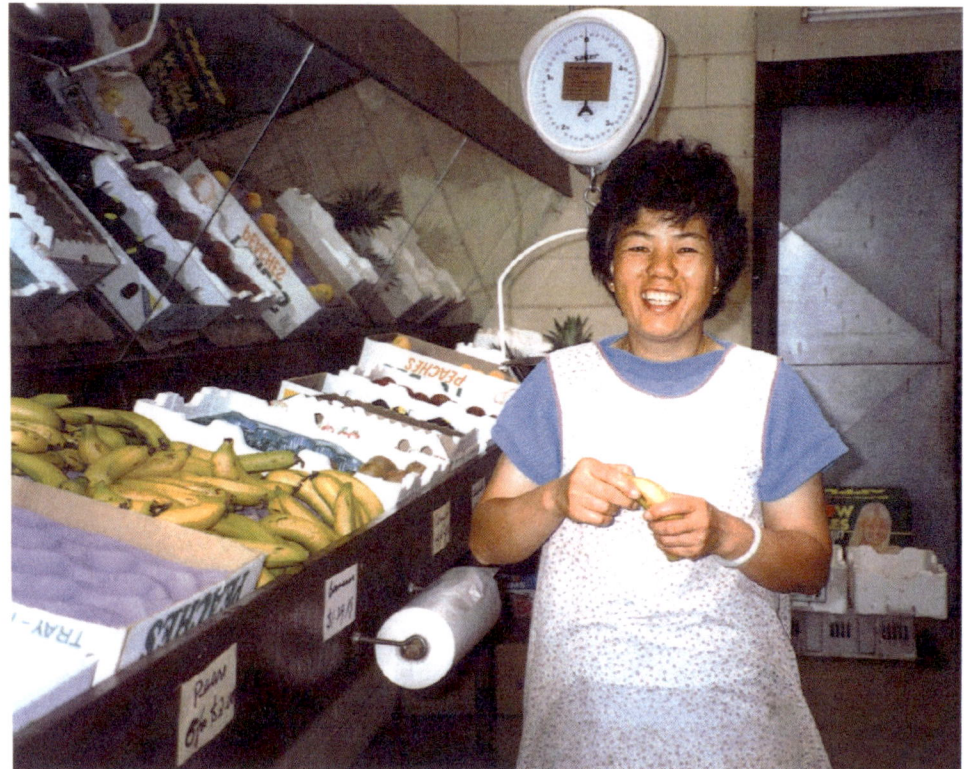

Figure 42 - Sim Wong in Hop Chong Garden Shop, Camden, 1980s (Wong family).

Figure 43 - Hop Chong Garden Shop, 3 Argyle Street, Camden 1990s (Wong family).

13 • Biu Wong

The Last Camden Chinese Market Gardener

Julie Wrigley

Biu Wong helped provide information about his life to complete the story of Camden's Chinese market gardeners.

Cheuk Biu Wong was born in 1940 in Canton, now known as Guangzhou, China, the son of Chinese market gardener Yung Sum Wong and Siu Mee Wong. Biu grew up in China with his mother, while his father, Yung Sum Wong, came to Australia to work and send money back to the family. Yung Sum Wong came via Port Vila in Vanuatu to Camden in about 1945. He joined the Hop Chong Garden group.

In May 1957, the Hop Chong group: Yung Sum Wong, Ho Jo Ling, Leung Ah Sin, Lo Lin Chow, and Wong Tim purchased land near the entrance to Camden on the western bank of the Nepean River from Henry Nelson Whiteman and Frederick Keith Whiteman [68]. Hop Chong Garden applied for a licence under the Water Act for a pump on the Nepean River for irrigation of twenty-three acres. [69]

Figure 44 - Biu Wong, with George Fung, Sim Wong's brother-in-law, who was visiting from Hong Kong, 1986 (Photo supplied by the Wong family to Camden Museum).

---

[68] Document held by Biu Wong and Sim Wong, purchase of 22 acres 2 roods and 2 perches.
[69] *Government Gazette* 24 May 1957 Water Act 1912-1955, on part portion 12 in the Parish of Camden in the County of Camden, p 1665.

13 • Biu Wong – The Last Camden Chinese Market Gardner

Biu Wong came to Camden from China when he was twenty-one and joined Hop Chong Garden with his father. Biu was keen to improve his English, so he applied to attend Camden High School as a mature-aged student. In 1963, he was accepted and attended for two years, but he needed help with English. He coped well with maths taught by Don Baker, who became a close friend, and later principal of Picton High School. As part of a government program, Nolene Baker, Don's wife, gave English lessons to Biu for one hour a week.

Biu Wong applied for Lau Sim Chan (Sim) to come to Australia from Hong Kong, where she had met Biu. Sim arrived in Australia in January 1965 and was naturalised in August 1967. [70]

Figure 45 - Wedding of Biu and Sim Wong. St John's Church, Camden, 1965 (Wong family).

Rev Kirk married them at St. John's Church, Camden, in 1965. The wedding was attended by John Mack, the town clerk and his wife Gloria Mack and son Greg, the parents of Biu Wong, family friends, St. John's parishioners, Don Baker, Nolene Baker and the Baker children Sally and Tom. The photo shows how Biu and Sim Wong forged positive relationships with the Camden community.

Biu Wong worked in the Hop Chong market garden and eventually purchased the land with his father's help. Biu organised the business differently from his father, who used draught horses to plough the fields. Biu introduced modern farming techniques, by purchasing a second-hand tractor, and introducing pipe irrigation. Biu's wife, Sim Wong, started selling vegetables by the side of the road in 1969. They followed the example of another stall selling vegetables on a busy road and decided to try it themselves, as there was little money selling vegetables wholesale.

---

[70] *Government Gazette* 3 August, Wong, Lau Sim.

13 • Biu Wong – The Last Camden Chinese Market Gardner

Figure 46 - Sim and Biu Wong inside the Hop Chong Garden Shop, 1980s (Wong family).

At the end of 1969, Biu built the Hop Chong Garden Shop, where his wife Sim and his children worked selling fruit and vegetables over the years.

Biu and Sim had five daughters, Anne, Dianne, Julie, Margaret, and Sally and one son, Andrew. The children attended Camden High School or later Elderslie High School and were all hard-working and well-liked. At Camden High School, Anne, Dianne, and Julie all won awards for English. In 1986, Dianne Wong was the school captain and dux of the school. [71] All of the children went on to attain university degrees. Biu was keen for all his children to gain a good education so they did not have to work as physically hard as he had.

Biu's father, Yung Sum Wong, died in 1972 and was buried in the Chinese section at Rookwood, with three generations so far following him in Australia. Biu's mother lived with the family for many years until she died in 2007 and was buried at Rookwood.

Biu Wong sold the farm in 1993 and remaining living in Camden. He recalls hard times, such as when his first tractor was stolen, other petty theft of his property, and the devastating flood of 1975, which washed away the previous Cowpastures Bridge. Additionally, the 1978 flood which washed away the previous Hop Chong Garden shop, and also, the huge flood of 1988. He said the life of a market gardener was a life of hard labour and long hours. He has been very generous to the Camden Museum to give back to the town that has supported him and his family.

---

[71] Camden High School 1956-2016, Alted Printing, Camden, 2016.

13 • Biu Wong – The Last Camden Chinese Market Gardner

Figure 47 - Biu Wong on his Massey Ferguson tractor, Hop Chong Garden, 1980s (Wong family).

Figure 48 - Sim Wong and Andrew on the same Massey Ferguson tractor, 1980s (Wong family).

14 • Artefacts in the Camden Museum Owned by the Chinese Market Gardeners

John Wrigley

Figure 49 - Wong family visiting the Nixon Room at the Camden Museum, 2022 (Andrew Lui).

Residents of Camden have donated items to do with the Chinese market gardeners to the Camden Museum. Some of these were dug up or found after the Chinese market gardeners had left the area. These items are significant in conveying something of the Chinese market gardeners' way of living in simple huts far from home.

Abacus

An abacus is a counting frame for addition, subtraction, division, and multiplication. It can also calculate a number's square and cubic roots. The beads are manipulated with the help of one hand's index finger or thumb. This abacus was used in Camden by the Chinese market gardeners who were tenants of Davies' farm from 1921-1963. It was found in the old house on the farm previously occupied by the Chinese market gardeners on the Davies property in Exeter Street. This item was a favourite of the donor, Miss Llewella Davies, who enjoyed demonstrating its use to young Museum visitors. Some Chinese gardeners could not read English but were extremely fast at their abacus calculations.

Figure 50 - Abacus. Item 1980.1 (Anne McIntosh).

Chinese Hat

The 'coolie' hat is made of straw or rattan with a string under the jaw tie, 45 x 45 x 13 cm. It dates from the 1940s or 50s and was donated by the Carmagnola family. The hat was used by Chinese market gardeners in Elderslie on the garden under the present Macarthur Bridge. Another straw or rattan hat with a black plastic chinstrap was owned and used by Mr. Biu Wong, the last Chinese market gardener in Camden. 37 x 33 x 14 cm. The Chinese market gardeners worked long days in all weathers.

Figure 51 - Chinese hat. Item 2012.24 (Anne McIntosh).

14 • Artefacts in the Camden Museum owned by the Chinese Market Gardeners

Chinese Silk Embroidery

The silk embroidery has the scene of a bird and tree blossom on a beige background in a gold-painted frame. 48 x 35 x 5 cm. The frame was repaired in 2022 after the string for hanging the frame broke. In Miss Llewella Davies' handwriting on the back of the frame is written: From Tong Hing Market Gardens on Exeter farm, Macquarie Grove Road 1914-1955. Loaned by L. Davies, Exeter Street Camden.

The Tong Hing Chinese market gardens were located on the Davies' family land in Exeter Street and Macquarie Grove Road Camden. From time to time the Chinese men brought back from China gifts such as jars of ginger or lychees for the families of the owners of the land which they leased. The silk embroidery was such a gift.

Figure 52 - Framed silk embroidery. Item 1980.283 (Anne McIntosh).

China Bowl

The bowl has a hand-painted scene of dragons in brown and gold. It is chipped and cracked, with Chinese markings on the base, 12 cm in diameter. The donor, Mr. Stan Aliprandi, was a trained pharmacist, a vigneron and lessee of Biu Wong's former market garden, Hop Chong (northwest of Cowpastures Bridge), where he developed a vineyard and found this item. It was made in China's 'Kiangsi' province, which is famous for making china and porcelain and is the main area in China for the production of fine china ware. This fine bowl is a contrast to the collection of brown earthenware bowls and jars in the Museum.

Figure 53 - Dragon bowl. Item 1995.449 (Anne McIntosh).

14 • Artefacts in the Camden Museum owned by the Chinese Market Gardeners

Rice Wine Bottles

The donor, Ngaire Thorn, dug up this item about 1960 on her parents' property in soil on the flood plain below *Caernarvon*, Macquarie Grove, beside the Nepean River, an area previously used by Sheil's Chinese market gardeners, trading as Sun Chong Kee. This item is a relic of that time and is like another bottle, Museum Item 1980.186. The bottles would have been imported from China with their contents and a cork in the top. Rice wine is powerful in alcohol, equivalent to whisky.

The Museum has a collection of brown earthenware eating bowls and jars which would have been exported from China. Narrow-necked jars would have held soy sauce or other liquids. The wide-mouthed jars would have come containing food and been reused as cooking pots. The jars represent the simple lifestyle the men endured away from their families, language, and culture.

From the 1970s to the present, residents of Camden have thought it important to remember at the Museum the Chinese market gardeners. They experienced hard times and significantly contributed to working the river flats along the Nepean River and making fresh vegetables available for Camden and Sydney.

Figure 54 - Rice Wine bottles. Items 2015.12 and 1980.186 (Anne McIntosh).

14 • Artefacts in the Camden Museum owned by the Chinese Market Gardeners

Figure 55 - Wong family in the Chinese bay of the Camden Museum, 2022 (Andrew Liu).

Figure 56 - Wong family in the Research Room at the Camden Museum, 2022 (Andrew Lui).

15 • Looking Back on the Camden Chinese Market Gardeners

Julie Wrigley

Today, visitors to Camden Museum are surprised to find that from the early 1900s to 1993, there was a community of Chinese market gardeners in Camden. Most of the names of the Chinese market gardeners have been forgotten.

However, many residents remember the last of the Camden Chinese market gardeners, Biu Wong. He was well known and respected by those who bought vegetables in the Hop Chong Garden Shop at the entrance to Camden. As previously explained, Biu retired in 1993, and Biu and Sim remain living in Camden. Biu Wong has donated generous photos and artefacts to the Camden Museum, and the Wong family has made an outstanding contribution to Australian life.

Other descendants of the Chinese market gardeners include Daphne Lowe Kelley, President of the Chinese Australian Historical Society based in Haymarket, New South Wales. Daphne has made a significant contribution to Australian life and believes we all should learn more about Chinese history in Australia.

Descendants

We were hoping more descendants of the Camden Chinese market gardeners would come forward, and this happened recently. In December 2022, Jackson Pan, a great-grandson of market gardener Foon Kee Pan (who worked at the Davies' garden in Camden), came into the Camden Museum.

Then, in June 2023, Foon Kee Pan's son, Ben Pan, contacted the Museum and provided details of his father, Foon Kee Pan and Uncle Yen Hung Pan. They were both Chinese market gardeners who worked on the Chinese vegetable garden at Macquarie Grove on the site now used for the Miss Llewella Davies Pioneer Walkway. Their names and signatures were on the Davies' lease in 1946 between Mrs. Mary F Davies and the Tong Hing syndicate (see Figure 11): FOON KEE [PAN]; HO JO LING, JACK YOUNG, LAW SANG, AH SEE, SOLOMON FUNGNAM; and YEN HUNG PAN.

Recent research using the National Archives of Australia has revealed at least four of these seven Chinese market gardeners working on Miss Davies' farm in Camden in 1946 had been evacuated from the British Solomon Islands Protectorate in 1942 when the Islands were invaded by Japan during World War II. Before contact with Ben Pan, we did not know that any Chinese market gardeners had come to Camden via the Solomon Islands.

YEN HUNG PAN arrived in Sydney per SS *Morinda* in January 1942 (National Archives of Australia SP11/2, Chinese/Pan Y H).

FOON KEE PAN came to Sydney on 17 February 1942 on SS *Morinda* (NAA SP11/2, Chinese/Foon K B) from Tulagi, a small island in the Solomon Islands. (Foon Kee Pan's wife, daughter and son, Ben Pan, were evacuated to Sydney in February 1942 on the SS *Bullolo*.)

SOLOMON FUNGNAM is recorded in the archives as NAM FUNG (Chinese), arrived in Sydney per SS *Morinda* on 17 February 1942, an evacuee from Tulagi, part of the Solomon Islands

15 • Looking Back on the Chinese Market Gardeners

(NAA SP 11/2, Chinese/Fung N). He came from the same small island as Foon Kee Pan and was evacuated to Australia on the same steamer.

HO JO LING came to Sydney on 12 June 1942 on SS *Morinda* (NAA SP11/2, Chinese/ Ho Jo Ling). See memories from Joy Riley about Foon Kee Pan and Ho Jo Ling, Chapter 5.

Figure 57 - Ho Jo Ling, Foon Kee Pan and Ben Pan as a boy, c. 1947.

The archives show the Department of Immigration authorities were concerned about whether or not the Chinese evacuees from the British Solomon Islands were eligible to secure employment whilst in Australia. On 11 April 1942, the Department of the Interior sent letters to the Collector of Customs, the Consul-General for China, and the Secretary of the Department of Labour and National Service stating:

> I am to inform you that no objection will be raised to the above-named Chinese taking up suitable employment whilst they are in Australia for the duration of the war, provided such employment will not be detrimental to Australian workers and on the understanding that it will be subject to the rates of pay and conditions of employment ruling in the districts in which they Chinese are employed. (NAA: A445, 236/2/14)

When he was in the Solomon Islands, Ben's father, Foon Kee Pan, had been a tailor, trained in Hong Kong, but was now permitted to work as a market gardener in Camden, in 'employment which will not be detrimental to Australian workers' – and work which was not desired by many Australian workers.

Ben Pan (now eighty-six years old) told me that the rest of the Pan family group were evacuated on other boats from the Solomon Islands, with some of the family delayed because of the war. The authorities confused the Chinese names and regarded Yen as the surname, so the extended family ended up frustratingly having different surnames of Yen and Pan.

Ben explained that he lived in Sydney with his mother and family but visited the Camden farm as a boy. Ben travelled with his father, Foon Kee Pan, on weekends and school holidays when he took produce to the Sydney market on the train, Pansy. Ben remembers that he went with his father to pay the rent to Miss Davies for the lease and that he swam in wells on the farm.

Ben said his father did not have time to spend with him because he worked long hours. Ben recalls the large shed where the workers lived and cooked, with a dormitory on one side, storage on the other, a verandah, and a place for the workhorses. He said the cooperatives did not have much contact with each other, but in times of heavy rain, they gathered and played Mahjong the whole time.

Ben said the Pan family were happy to settle in Australia. His father, Foon Kee Pan, did not return to China as he thought it might be difficult to return to Australia. Foon Kee Pan worked on the farm until the 1950s, when he had a fall and could not work. He left Camden and died in Sydney in 1955. Ben Pan is now 85, but his sons carry the Pan name.

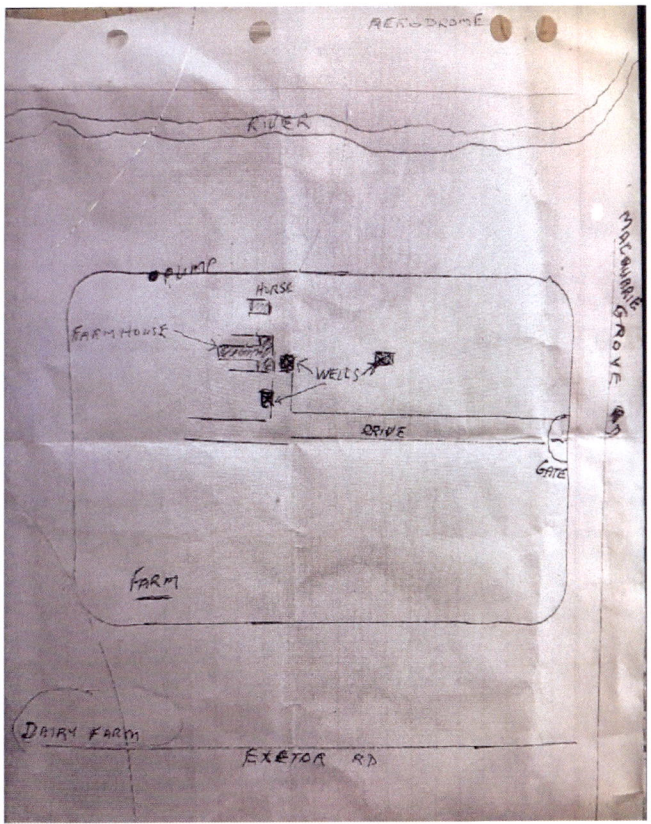

Figure 58 - Sketch of Tong Hing market garden on Davies' farm, hand-drawn by Ben Pan, 2023.

15 • Looking Back on the Chinese Market Gardeners

Another descendant was uncovered in July 2023. Charlie Louie came to the Museum and showed photos of his grandfather, Louie Yiyou, who delivered vegetables to Haymarket Farmers Market from Camden in the 1920s. His grandfather returned to China and died in 1925 in Guangdong. This book might lead to uncovering some more descendants.

Visitors to the Camden Museum are pleased to find that the contribution of the Chinese market gardeners is honoured. Their farming tools and the collection of their modest eating bowls show the hardship of their day-to-day lives. Some of these eating bowls have been dug up and donated to the Museum by residents who want to remember the local market gardeners. The late Llewella Davies OAM and Bruno Carmagnola are Camden residents who have greatly contributed to remembering the market gardeners. Camden Town Farm on Exeter Street was originally a 52-hectare dairy farm that was bequeathed to the local community by Llewella Davies. The farm is owned by Camden Council and managed by a Council Community Committee of dedicated local community volunteers. One site on Miss Davies' Walk on the farmland shows where the Chinese gardeners dug wells for their gardens.

Some items in the Museum are not to do with the Chinese market gardeners but reminders of Chinese culture and the delight in 'Chinois', a style inspired by art and design from China, Japan, and other Asian countries.

A beautiful hand-painted cup and saucer with Chinese floral decoration dating from the 1880s inspired Chinese filmmaker and resident Wen Denaro to make a film, *100 Yards of Silk,* about the Chinese in Camden. This interesting twelve-minute film can be viewed at the Museum.

There are records of the Chinese market gardeners on Trove, the National Archives of Australia, and the State records at Kingswood [72], and there are two names, Willie Chung and Wong Yong, who are life members of the Camden District Hospital Honour Board.

However, in Camden, the Chinese market gardeners have gone and have been forgotten by many. The Museum remembers their contribution to farming and settlement in the Camden area of New South Wales.

---

[72] State Archives and Records Authority (NSW), 161 O'Connell Street, Kingswood. 2747.

15 • Looking Back on the Chinese Market Gardeners

Figure 59 - Wong family with John and Julie Wrigley at Camden Museum, 2022 (Andrew Lui).

Figure 60 - Biu Wong, Sim Wong and John Wrigley at the Camden Museum, 2022 (Andrew Lui).

Conclusion

Ian Willis

The story of Chinese market gardeners in the Camden area has been largely overlooked, and there is little tangible evidence of their presence in the area from the past. Yet a little digging reveals a rich and complex history that made a significant contribution to the local economy, interracial relations and agricultural innovation and technology.

The country press is a valuable and dense source of resources for researching rural life in Australia. Yet, the Camden press provides little contemporary coverage of the farming activities of Camden Chinese market gardeners. The formation of the Camden Historical Society in 1957 was primarily concerned with losing European history with the flooding of the Burragorang Valley. It was mostly silent on the Chinese story. And unfortunately, the absence of a local joss house, temple, or Chinese store has only added to the silence.

The telling of the story of the Chinese market gardeners only really commenced with the opening of the Camden Museum in 1970 by the historical society. Several objects were donated to the Museum by Southern European horticulturalists who took over what where were previously Chinese market gardens. The objects were recovered relics of a lost past, and their public display became central to telling the story of the Chinese market gardeners. (see earlier chapters). This process was assisted by the work of local sage RE 'Dick' Nixon and, in more recent times, J&J Wrigley. The Wrigleys have assisted members of the Chinese community in telling their stories through filmmaking and oral and family history.

This publication has attempted to fill in some gaps in the Camden Chinese market gardeners' story. At the same time, other researchers have done work on the presence of Chinese market gardeners elsewhere in Australia. [Boileau 2017; Wilton 2004; McGowan 2010; Kwok 2019]

The Floodplain

The Camden Chinese market gardeners first settled on the Nepean River floodplain in and around the township of Camden in the late 1890s. Camden had been established in 1840 as a private venture by the Macarthur family on the Great South Road around 60 km southwest of Sydney. It was located at the river crossing and entry to the family's Camden Park Estate. The railway and the Great South Road crossed the floodplain and entered Camden at the Nepean River crossing.

The site and situation of the Nepean River floodplain at Camden provided the land that was essential for the Chinese market gardeners to practise horticulture. The floodplain was an attractive area with rich pastures that had first drawn the escaped cattle of the First Fleet in 1788 and named the Cowpastures by Governor Hunter in 1795.

The land was central to the identity of the displaced occupants, the Dharawal, the European occupiers and then the Chinese market gardeners. European colonisation dispossessed the Indigenous occupants, and neither the Chinese nor Europeans had much time for the area's Indigenous people. The Camden Chinese made no effort to counter the forces of dispossession or displacement of First Australians.

Conclusion

European and Chinese settlers used the land to legitimise their place in the world, assert their identity and lay claim to the territory. [Boileau:235] Boileau maintains that the Chinese use of land sheds light on the cultural issues of power relations, colonisation, work and leisure, and subsistence. She argues that gardens and farming fields carry symbolic, cultural, and social significance attributes as constructed landscapes and bounded spaces. [Boileau:235]

The identity of Chinese and Europeans alike was bound up in working the land, and Boileau argues that the Chinese succeeded as yeoman farmers, whereas the Europeans failed. [Boileau:20] The Chinese were part of the pioneer story yet have never been acknowledged for their efforts in taming the land through horticulture.

By the time the Chinese market gardeners arrived on the Nepean River floodplain in the 1890s, most of the native vegetation, which consisted of Sydney Coastal River Flat Forest and freshwater wetlands (Pittendrigh Shinkfield Bruce 2002), had been cleared by Europeans for grazing and cropping. The settler colonial project, which had commenced with the European occupation of New South Wales, readily absorbed the horticultural activities of the Chinese market gardeners alongside European farming.

The farming community along the floodplain in the early twentieth century was dominated by dairying activities of Camden Park Estate between Menangle village and Camden township on the left bank of the river. The Chinese market gardeners located themselves on the left bank at Camden and the right bank from Elderslie to Cobbitty village. Restrictions on land ownership meant that the Chinese leased land from local European landowners, as in other locations like Bathurst, where share-farming was practised. (Kwok 2019: 215, 231) One of Camden's dairy farmers, the Davies family, kept rental payments by Chinese market gardeners in their rent receipt book (illustrated earlier).

Relatively dry winters on the floodplain meant the Chinese market gardeners needed to rely on irrigation from the Nepean River to water their crops. The Chinese used flood irrigation by pumping river water onto their field between the cropping furrows. They used large-bore belt-driven pumps using movable steam engines, which were easily repaired after flooding. Later, the use of gas and electric-driven pumps made the Chinese more reliant on the town's blacksmiths and electricians. [Nixon] Similar to Bathurst's Chinese market gardeners who switched to electric pumps after electricity was introduced to the town in 1925. (Kwok 2019: 254)

The Chinese market gardeners' choice of crops was guided by the area's warm to hot summers with cool to mild winters with regular frosts. Winter crops included cabbages, cauliflower, turnips, sprouts, and other brassica crops. Summer crops included tomatoes and beans. [Nixon] Crops yields were impressive.

Field preparation for cropping was initially done with hand tools and horses, which included hoeing, planting, fertilising, weeding, and harvesting. Similarly, most European dairy farmers relied on hand milking and transport by horse and cart. After the Second World War, the Chinese used rotary hoes and tractors in their fields.

Innovative Self-employed Smallholder Farmers

Chinese and Europeans were self-employed smallholder farmers whose aspirations hoped for success and status revolved around the farms making a profit. The Chinese, like Camden's European farmers, were entrepreneurs or risk takers, which meant they occasionally turned a loss due to flooding and the resulting destruction of crops, as well as loss of equipment and housing.

Conclusion

The Chinese market gardeners undertook horticulture as a labour-intensive, low-capital-input farming activity that was not popular with European farmers. The Chinese market gardeners did not directly compete with other Camden's European smallholder dairy farmers and orchardists. Camden European town dwellers and smallholder farmers considered horticulture a low status activity. They only grew fruit and vegetables in an unorganised, informal manner for personal use in their 'home' gardens.

Camden European and Chinese farmers used their agency, as Boileau calls it, to follow their aspirations and sought to achieve their own goals [Boileau:268]. Both groups of farmers were active agents of history as innovators and risk takers, developing flexible coping strategies and using their technological skills to their advantage [Boileau: 62]. Both displayed their produce at the Camden Show, won prizes and gained considerable kudos for their skills and effort. In contrast, Chinese market gardeners at Bathurst faced opposition and restrictions at the annual show at the turn of the twentieth century. Yet, by 1925 there was a section called 'Best Collection of Vegetables grown by a Chinaman'. (Kwok 2019: 221)

The Chinese market gardeners drew their farming adaptations from several influences, including a cultural context where they drew on Chinese traditions of subsistence agriculture in southern China. Similarly, Camden's European dairy farmers drew on global dairying innovations to increase their profits and productivity and new technology from the Northern Hemisphere. Trial and error were part of the farming modus operandi of both groups of farmers.

Warwick Frost argues that the Chinese in the late nineteenth century and early twentieth century dominated intensive agriculture in Australia and acted as agents of technological change. [Boileau:166-167] This certainly applied to the Camden Chinese market gardeners. [Nixon] The Camden Chinese worked in teams through class associations [Boileau:228] and reciprocal obligations within each work team with ascribed roles. [Nixon] They had networks and cooperative work arrangements that supported the growth of their market garden businesses. Their networks created efficient, informal support systems and a source of financial and social capital [Boileau:228-229] in Sydney.

Road and Rail Transport

The railway and Great South Road were transport routes that provided a thoroughfare for the movement of ideas, people, technology, and other global forces to the Camden floodplain in and out of the district. The Sydney markets were readily accessible by the railway and road for perishable horticultural produce from the Nepean River floodplain. European dairy farmers had used the railway since its arrival in 1881, and the Chinese readily saw its advantages. Rail wagons were loaded by hand with horticultural produce brought to the Camden yards by horse and cart.

Chinese marketing arrangements, as were those of European dairy farmers, were highly organised. Chinese work teams were highly structured, and the market man was allocated to supervising the sale of the produce at the Sydney market. As road transport became more reliable in the interwar years, the Camden Chinese used the services of European truck carriers to transport produce to the Sydney market.

The Chinese used the railway to bring sheep and horse manure to Camden from racing stables and the Flemington saleyards. When the manure arrived at Camden railway yards, it was unloaded by hand and carted by horse-pulled dray to the market gardens. Manpower was then used to spread manure across the fields on the floodplain. [Nixon] The Chinese were subject to the vagaries of the market and fluctuating prices for their produce. In contrast, European dairy produce had a degree of price control and stability.

Between Two Worlds

The Chinese were 'The Other' in the Camden story. They were outsiders and always remained foreigners. The Camden Chinese lived between two worlds [Boileau:130], whereas the Camden European smallholder dairy farmers were part of the area's dominant culture. The Chinese conducted their farming in a hostile environment created by immigration restrictions and the White Australian Policy. When any of the Camden Chinese wanted to return to China, they faced immigration restrictions on their return to Australia.

The Camden Chinese were part of local business networks, and their contact with Camden Europeans was primarily transactional as participants in the local marketplace. While the Camden Chinese were well-respected citizens, they resisted assimilationist pressures [Nixon] and 'kept to themselves'. They spoke one of the many Cantonese dialects and had their diet, food requirements, and religious and cultural practices.

The Chinese and European farming communities relied on the town for services and other activities. The Camden township and surrounding villages provided a secondary market for the Chinese who sold door-to-door, as did some European dairy farmers. [Boileau:140-141] The township and its socially tight-knit European community were also a source of supplies for the Chinese.

The Camden Chinese lived a frugal communal lifestyle as single men [Boileau:137], whereas Camden's European smallholder farmers were family-based with a sense of permanency. The lack of Chinese women in their community did not allow them to establish family ties in Camden, unlike the European community, where women were critical to developing family and community networks. The Chinese community had a temporary and transitory nature, whereas the European community had a sense of permanency that allowed asset and capital accumulation.

Paradoxically, the Camden community both shunned and embraced the Chinese market gardeners. Europeans in Camden empathised with local Chinese in the face of natural disasters like the regular flooding from the Nepean River and purchased their produce daily. On the other hand, the European community did not encourage Chinese participation in the town's social and cultural activities. Camden Europeans made little or no effort to learn their Chinese names and resorted to slang terms or used the term Chinamen to identify them as a group. [Nixon]

Parochialism or local patriotism was important to Camden's community identity and sense of place. Chinese philanthropy supported Camden's parochialism in several directions and was a vehicle that allowed the Chinese market gardeners a degree of inclusion by supporting good causes. The Chinese made donations and subscriptions that were well above those of the equivalent of smallholder European farmers. Chinese market gardeners were listed on the wartime patriotic funds subscription lists to demonstrate patriotism during the First and Second World Wars, similar to other minority groups. Yet, they never appeared on the organising committees. In 1943, the Camden Women's Voluntary Services organised fundraising, with no Chinese input, for the China Day Appeal, raising £72/17/3. Donors were not listed, and the WVS reported that 'the greatest effort could not be too much effort for such a gallant ally'. (*Camden News*, 10 August 1944) While at Bathurst, the European and Chinese communities combined to raise £500 for the appeal. (Kwok 2019: 266) When the Camden Chinese received medical treatment, they donated to the Camden District Hospital.

Conclusion

The Chinese market gardeners had a broad view of their world that ranged from their homeland in China to Camden but never saw themselves as part of the British Empire. They networked with the Sydney Chinese diaspora, who provided them with language services, financial services, personal contact networks, letter writing and other facilities. Both Camden Europeans and Chinese kept in contact with their homeland. The Camden Chinese intended to die in China, not Australia, made regular visits home and sent letters to family in China. Similarly, Camden Europeans travelled home to Europe and sent letters, money and goods to friends and family. Chinese used chain migration as did other immigrant groups from Southern Europe in the mid-twentieth century [Boileau:131]

The Legacy of the Camden Chinese Market Gardeners

The Camden Chinese market gardeners added a rich layer of history to the story of a small country town. The Chinese in the Camden area brought their culture, farming and marketing skills, entrepreneurship, inventiveness, adaptability, and resourcefulness. Through innovation and persistence, they succeeded at a type of farming that the local European farmers practised in a small, unorganised, informal fashion in their 'home' gardens in and out of the township and villages.

The Camden Chinese market gardeners were upstanding citizens who contributed to local society and the economy. They paid their debts and met their financial obligations. The Camden Chinese market gardeners supported local causes, including Camden Hospital and local patriotic funds during wartime. The Chinese provided a stable rural labour force when the emerging Burragorang coalfields in the 1930s started attracting European labour from local farms.

The Camden Chinese market gardeners enriched the community through ongoing interracial relations. The Chinese employed local Europeans to pick, cart, transport produce, plough their fields and repair and maintain irrigation equipment. The Chinese used the medical facilities at the Camden District Hospital and rented their farmland from local landowners. During the regular floods on the Nepean River, the Chinese received shelter and sustenance from the Camden European community. The European community bought Chinese produce and sold them supplies.

The Camden Chinese were successful [Boileau: 150-151] creative entrepreneurs and innovators in a hostile, challenging environment, particularly after Federation. They enriched the local economy through innovation of farming practices, irrigation technology, and marketing. The Chinese were successful smallholder farmers and businessmen who engaged with the local community and significantly contributed to the local economy.

For the first time, Camden Chinese market gardeners established intensive horticulture on the Nepean River floodplain. They demonstrated this was a viable and profitable form of agriculture using mechanically assisted flood irrigation. The Camden Chinese market gardeners practised a form of labour-intensive agriculture that produced high yields per acre on small farms. They were efficient and effective innovators, hard-working, and occupied a niche market in the Camden agricultural economy.

The Chinese market gardeners were adaptable, innovative, and resilient in the face of many challenges, particularly flooding on the Nepean River floodplain. Their pioneering efforts in horticulture are still replicated on the same sites on the floodplain by descendants of Southern European immigrants who largely replaced the Camden Chinese in the 1950s. One of these families was the Vellas of Maltese descent. In 2021, the third generation of the Vella family opened a lettuce and cauliflower farm on the right bank of the Nepean River floodplain near the former site of Sun Chong Kee's market garden. Matt Vella said, 'His family had been farming in Camden and

Conclusion

Wollondilly for over 50 years. His grandfather was a vegetable farmer in Malta (*Camden Advertiser*, 26 May 2017)'.

The Vellas, who operated two other farms at Ellis Lane on the left bank of the Nepean River, said their first harvest was beyond expectation. Matt Vella stated that the land 'was ideal for farming', which the Chinese market gardeners had found over 50 years before. Like the Camden Chinese, Vellas shipped the produce to the Sydney markets for sale. (*Campbelltown Macarthur Advertiser*, 8 December 2017) Interestingly, neither story in the Macarthur Press referenced the heritage of the Chinese market gardeners or their farming activities on the floodplain.

The Camden Chinese market gardener's legacy is evident in many parts of the district, much hidden in plain sight. The history of the Camden Chinese market gardeners is a rich area for further research, and there is still much to tell in this story.

Appendix

Chinese Deaths in Camden

Some of the Chinese deaths recorded in Camden between 1920 and 1970 can be found on the NSW Births, Deaths, and Marriages Register [73]

- CHUN YUEN — Died 1921, aged 51
- AH CHONG — Died 1922, aged 62
- CHONG HING — Died 1925
- LOO HIM WONG — Died 1935
- GARN ON — Died 1935, aged 59
- LOWE KUM MING — Died 1935, aged 55
- LOO HIM WONG — Died 1935
- WING LEE — Died 1938, aged 65
- WON WAH — Died 1938, aged 45
- CHONG SING — Died 1939, aged 70
- KUM GEE — Died 1939, aged 58
- GEORGE CHONG — Died 1940, aged 73
- JOE YOUNG FOO — Died 1940, aged 64
- PANG FOOK — Died 1940, aged 68
- KUM HOW — Died 1941
- JOE LONG — Died 1943, aged 72
- LEE GUM — Died 1944
- AH KONG — Died 1945
- HONG TONG — Died 1945, aged 64
- WONG QUAY — Died 1949
- TUNG SING — Died 1951, aged 70
- CHUNG SING — Died 1955, aged 70
- YUEN THONG — Died 1956, aged 45
- YEW GEE — Died 1956, aged 78
- SOW WONG — Died 1966, aged 86

---

[73] Births, Deaths, and Marriages NSW: https://familyhistory.bdm.nsw.gov.au

Bibliography

Boileau, Joanna 2017, *Chinese Market Gardening in Australia and New Zealand, Gardens of Prosperity*. Palgrave Macmillan, Cham, Switzerland.

Camden High School. 2016, *Camden High School 1956-2016*, Camden High School, Camden.

Jones, Doris Yau-Chong. 2003, *Remembering the forgotten: Chinese gravestones in Rookwood Cemetery 1917-1949*, Doris Yau-Chong Jones Invenet Pty. Ltd. Publishing Pymble, NSW

Kwok, J. 2019. The Chinese in Bathurst: Recovering Forgotten Histories. [Doctoral Thesis, Charles Sturt University]. Charles Sturt University.

McGowan, Barry 2010, *Tracking the Dragon, A history of the Chinese in the Riverina*. Wagga Wagga Museum of the Riverina and Powerhouse Publishing, Sydney.

Pittendrigh Shinkfield Bruce 2002, *Camden Riparian Areas Plan of Management*. Camden Council, Oran Park.

Stuckey, Frank. 1987, *Our Daily Bread, The story of the Stuckey Bros, Bakers and Pastrycooks of Camden NSW, 1912-1960*. Camden Historical Society, Camden.

Williams, Michael. 1999, Chinese Settlement in NSW, a thematic history: a report for the NSW Heritage Office of NSW. State Library of NSW

Willis, Ian. 2015, *Pictorial History of Camden & District*. Kingsclear Books, Alexandria.

Wilton, Janis. 2004, *Golden Threads, the Chinese in Regional New South Wales 1850-1950*. New England Art Museum & Powerhouse Publishing, Sydney.

Yen, Mavis Gock. & Horsburgh, Richard. & Yen, Siaoman, 2022, *South Flows the Pearl: Chinese Australian Voices* / Mavis Gock Yen; edited by Siaoman Yen and Richard Horsburgh, Sydney University Press [Sydney, NSW]

Acknowledgements

Thanks to those who donated items to the Camden Museum to preserve the record of the Chinese market gardeners in Camden. Miss Llewella Davies OAM and Richard Nixon OAM contributed many items in the past. The gift from Miss Davies to the town for the Pioneer Walkway allows us to walk where the Tong Hing Davies' market gardeners worked. Bruno Carmagnola has donated farming artefacts to the Museum. Joy Riley has helped with photographs and memories.

Daphne Low-Kelley has helped with information and a photo of her grandfather, Chung Wong Ying (Wong Tong). Ben Pan has been a great help in adding to the story of his father, Foon Kee Pan.

The generous contribution of the Wong family, particularly Biu and Sim Wong and Sally and Andrew Liu, has been outstanding. They have contributed objects, photographs, and knowledge.

Thanks to Anne McIntosh for photographing Museum objects and Rene Rem for patiently scanning photos at a higher resolution.

Thanks to Sally Quinnell, our local member for Camden, for writing the foreword.

Thanks to Dr. Sophie Loy-Wilson for initiating the idea of a book about the Chinese market gardeners. Thanks to the Camden Historical Society for encouraging the project and to Fletcher Joss for preparing the document for printing.

Author Biographies

Sophie Loy-Wilson

Sophie Loy-Wilson, BA (Hons) PhD Sydney, Senior Lecturer in Australian History, is a historian of Chinese Australian communities. Her first book was a study of China-Australia relations in the interwar years, seen through the prism of Chinese Australian communities in Shanghai. In the book, she highlighted the importance of economic archives for immigration historians; these archives often preserve migrant agency. She combines methodological insights from labour history, overseas Chinese history, and the New History of Capitalism to bring a 'New Materialist' approach to Australia's multi-ethnic and multi-lingual past.

R.E. Nixon

Richard Edward Nixon (1919-2008) was born and died in Camden and had a lifelong interest in history. When the Camden Museum was extended in 1980, the upstairs exhibition chamber was named The Nixon Room in his honour and contained many items he collected for the Museum. Richard was president of the Camden Historical Society from 1991 until 1995. For his outstanding community service, he received many awards and honours. He was Camden's Citizen of the Year in 1990 and was awarded the Medal of the Order of Australia (OAM) in 1991. In 1976, Richard presented a paper explaining how the Chinese market gardener cooperatives operated and outlining the cultivation and irrigation methods the Chinese market gardeners used to produce vegetables for the local and Sydney markets.

Ian Willis

Dr. Ian Willis has been an honorary fellow at the University of Wollongong and completed his PhD in Australian History at Wollongong in 2004. He is president of the Camden Historical Society. He was awarded a Medal of the Order of Australia (OAM) in 2019 for his service to community history. His general area of research is centred on local studies in and around the Macarthur region of New South Wales, with works published in popular media to peer-reviewed journals.

Author Biographies

Julie Wrigley

Julie Wrigley née Gates completed her BA (1963) and Dip Ed (1964) at Sydney University. She taught English at Camden High School, St. Gregory's College Campbelltown, and St. Columba's High School Springwood. She retired from being English coordinator at St. Columba's High School in 2005. She and her husband, John, have lived in Camden since 1971. They have one daughter, Katie, who has one daughter Amber. In Camden, Julie is involved with the Camden Historical Society and volunteers at the Camden Museum. Julie writes history articles for the local newspaper and the Historical Society Journal.

John Wrigley

John Wrigley graduated with a Bachelor of Science (Forestry) from the Australian National University in 1965. John was a Sydney Water Board catchment area manager before he retired. He is a past president of the Camden Historical Society and vice president. In 2006, John received the Medal of the Order of Australia (OAM) for service to the Camden community. John has written books on Camden for the Camden Historical Society: *A History of Camden*, *Pioneers of Camden*, *Historic Buildings of Camden*, and *The Best of Back Then*.

Index

Newspapers

Camden Advertiser, 100

Camden News, 18, 41, 42, 47, 48, 51, 55, 56, 57, 58, 60, 65, 73, 75, 76, 98

Campbelltown Macarthur Advertiser, 100

Canberra Times, 60

Daily Advertiser – Wagga Wagga, 60

Daily Mercury – Mackay, 59

South Coast Times and Wollongong Argus, 18

The Daily Telegraph, 41, 60, 62

The District Reporter, xi

The Maitland Daily Mercury, 41

The Picton Post, 39, 41, 60

The Sun, 41, 65

The Sydney Morning Herald, 41, 48, 60, 61

Truth, 60

Other

100 Yards of Silk, xi, 62, 92

Camden Cottage Hospital Appeals, 75

Camden Council, 62, 92, 103

Camden Courthouse, 62

Camden District Hospital, 19, 42, 55, 65, 66, 67, 73, 74, 75, 76, 92, 98, 99

Camden Football Club, 30

Camden Historical Society, ii, ix, xi, xii, 14, 19, 33, 39, 40, 62, 70, 95, 103, 105, 107, 108

Camden Museum, ii, iii, xi, xii, 14, 17, 19, 31, 33, 37, 39, 40, 47, 49, 56, 59, 62, 70, 81, 83, 89, 92, 93, 95, 103, 105, 107, 108

Camden Show, 48, 97

Certificates Exempting from the Dictation Test, 17

Chinese Heritage Association of Australia, 66

Government Gazette, 75, 76, 79, 80

Immigration Restriction Act 1901, 17, 66

Museum of Chinese in Australia, 66

Nankin, 70

National Archives of Australia, xi, xii, 48, 49, 50, 51, 52, 53, 59, 60, 61, 67, 68, 70, 73, 89, 92

Pansy the Camden Tram, 19, 55, 57, 59, 91

RAAF, 38

Red Cross, 65

SS Bullolo, 89

SS Changsha, 68

SS Morinda, 89, 90

SS Tanda, 70

Sun Tong Hing Company, 37

Taiyuan, 51

Trove, xi, xii, 18, 92

Index

Water Act, 79

People

Adams, R.A.C., 37
Ah Chang, 76
Ah Chong, xi, 55, 56, 57, 101
Ah Hoe, 70, 73
Ah Kong, 76, 101
Ah Ling, 41
Ah Lum, 70, 73
Ah Lung, 56
Ah On, xi, 59, 60, 61, 62
Ah See, 37
Ah Sing, 56
Aliprandi, Mr. Stan, 86
Annand, Douglas, ii, 14, 38
Archbishop of Sydney, 47
Atkinson, Alan, 14
Baker,
   Don, 80
   Nolene, 80
Baldock, A.E., 51, 62
Baxter, George, 47, 48
Bean, Frank, 26
Bi Toa, 66
Big Charlie, 29
Biu Wong, xii, 20, 21, 29, 66, 74, 76, 79, 80, 81, 84, 86, 89, 93
Boardman, Allen, 61
Bond
   Brothers, 27, 44
   James Adam, 27, 39
Boodbury, 14
Brown, Mary Hilda, 67
Butler, Mr., 62
Carmagnola
   Bruno, 34, 35, 92, 105
   Maria, 34
Carmagnolas, 40
Charlie Chung, 67, 68, 69, 70
Charlie Louie, 92
Charlie the Chinaman, 29
Cheuk Biu Wong, 76, 79
Chine Leung, 51
Chong, Arthur, 62
Chong Hing, 101
Chong Lung, 76
Chun Yuen, iii, ix, xi, 51, 52, 53, 101
Chung Sing, 101
Chung Wong Ying, 66, 105
Cornhill, Richard, 40
Crane, H.W., 51
Crookston, Dr., 62
Davies
   Mary Fabert, 36, 37
   Miss, 36, 40, 44, 91, 92, 105
   Miss Llewella, 36, 44, 83, 85, 89, 92, 105
Dawson, Sergeant, 76
Dunk, Jack, 44, 45
Dunn, Mariam, xi, 42, 43
Evans, Christopher, 62
Fatt, Ernie, 76
Foon Kee, 37, 46, 89, 91, 105
Foon Kee Pan, 46, 89, 91, 105
Frost, Warick, 97
Fungnam, Solomon, 37
Gander, Albert and Allan, 27

Gao Yao, 75
Garn On, 101
Gates, Norman, 75
George Chong, 101
Hallam, Robert, 75
Hay Wing, 75
Haylock, Harry (Constable), 60
Hayter, Edward, 25
Ho Jo Ling, 37, 45, 46, 76, 79, 90
Hodge, Mr. Ben, 51
Holly Tom, 27
Hong Tong, 101
Hop Chong, ii, iii, xi, 20, 21, 29, 40, 42, 66, 70, 73, 74, 75, 76, 79, 80, 81, 86, 89
Inglis, Mr. R.H., 47
Jin Ming, 60
Joe Long, 101
Joe Young Foo, 101
Johnson, Edgar, 61
Joss, Fletcher, xii, 105
Keane, Joe, 26
Kelloway, H.S., 58
Kirk, Rev, 80
Ko Ming, 62
Kum Gee, 101
Kum How, 101
Kum Ming, xi, 59, 60, 61, 62, 63
Kwan Yin Tai, 76
Lai Wah, 41
Larkin, W, 47
Lau Sim Chan, 80
Law Sang, 37
Lee, George, iii, xi, 18, 47, 48, 49, 50
Lee Gum, 101

Lee Hing, 76
Leon/Leong Poy, 70
Leong/Loon Poy, 73
Leung Ah Sin, 76, 79
Little Charlie, 29
Lo Lin Chow, 76, 79
Locke, Mary, 39
Long
   Joe, 58
   Mrs. Annie, 68
Long Lee, 58
Loo Him Wong, 101
Louie Yiyou, 92
Lowe, Allen Wilfred, 57
Lowe Kum Ming, iii, xi, 59, 60, 62, 63, 101
Loy-Wilson, Sophie, xi, xii, 13, 14, 105, 107
Luen Fook Tong, 19
Macarthur-Onslow, Mrs., 48
Mack
   Gloria, 80
   John, 80
Marden, Jim, 27
Matherson, Alexander, 68
McCrae, Ian, 62
McIntosh, Anne, 36, 85, 87, 105
McKnight, Stan, 26
Mee King, 75
Moss, Herbert, 60
Neale, A.E., 41
Nesbitt, 21, 29, 43, 44
Nesbitts, 43
Nixon

Index

  Leslie, 25, 39
  Richard, xi, 21, 31, 34, 39, 40, 49, 70, 73, 105
On Chong Sing, 57
Pan,
  Ben, 89, 91, 105
  Jackson, 89
Pang Fook, 101
Paul, Rev T.G., 30
Peters, P, 51
Pitao, 66
Porteus, Sergeant E.H., 60
Price, Bert, 25, 72
Quinnell MP, Sally Anne, ix
Rapley, Billy, 26
Rawson
  Lady, 48
  Sir Harry, 48
Richardson
  Ern, 27
  Norman, 27
Ricketts, Charlie, 26
Riley, Joy, xi, 42, 43, 44, 46, 90, 105
San Yeck, 21, 43
Sheil, 21, 28, 29, 39, 60, 87
Silvio, 34
Sim Wong, 79, 80, 93, 105
Sing Mo, 21, 29, 40
Siu Mee Wong, 79
Southwell
  David, 42
  John, 42
Sow Wong, 101
Stuckey, Frank, 39, 40

Sue Ah Chong, 56
Sun Chong, 21, 28, 29, 40, 59, 60, 87, 100
Sun Chong Kee, 21, 28, 29, 40, 59, 87, 99
Sun Hop Chong, 42, 76
Sun Jing, 59
Taplin, William, 25
Tegel, Mr. AA, 40
Thorn, Ngaire, 87
Thornton, George, 26
Thurn, 21, 29, 34, 43, 44, 75
Tippetts, Archie, 27, 65
Tong Hing, 21, 26, 28, 29, 30, 36, 37, 40, 42, 85, 89, 105
Toong Goon, 70
Tung Sing, 101
Ung Lee, 51
Vella, Matt, 99, 100
W Chung, 73
Wen Denaro, xi, 62, 92
West, Doctor F.W., 51, 70
Whiteman, 21, 29, 79
  Henry Nelson, 79
Willie Chung, xi, 29, 65, 66, 67, 68, 70, 72, 73, 74, 76, 92
Willie Foote, 76
Willis, Ian, ix, xi, 95, 107
Wing Lee, 101
Won Wah, 101
Wong Quay, 101
Wong See, 56
Wong Tim, 76, 79
Wong Tong, 66, 73, 105
Wong Yong, iii, xi, 65, 66, 67, 92
Woods, Mrs. Jane, 68

Wrigley
  John & Julie, ix, xi, xii, 13, 14, 16, 17, 31, 33, 39, 40, 43, 47, 51, 55, 59, 63, 65, 67, 74, 75, 79, 81, 83, 89, 93, 95, 108
Yee Lee, 21, 25, 28, 29, 40, 42, 70
Yen Duck, 56
Yen Hung Pan, 37, 89
Yew Gee, 101
Yong Lee, 21, 29, 42
Young, Jack, 37
Young Lee, 21, 29, 40
Yung Sum Wong, 76, 79, 81

Places

AH&I Society Hall, 23, 24, 41
Alexandria, 19, 103
Argyle Street, 43, 44
Bathurst, 96, 97, 98, 103
Blooms Chemist, 44
Burragorang Valley, 95
Caernarvon, 39, 60, 87
Camden Cottage Hospital, 48, 75
Camden District Hospital, 19, 42, 55, 65, 66, 67, 73, 74, 75, 76, 92, 98, 99
Camden High School, 80, 81, 103, 108
Camden Museum, ii, iii, xi, xii, 14, 17, 19, 31, 33, 37, 39, 40, 47, 49, 56, 59, 62, 70, 81, 83, 89, 92, 93, 95, 103, 105, 107, 108
Camden Park Estate, 95, 96
Camden School of Arts, 40
Camden Valley Inn, 28
Camden Weir, 21
Canton, 17, 51, 55, 56, 59, 66, 70, 79
Carrington, 28
Cawdor Methodist Church, 28
Cowpastures, 47, 56, 75, 76, 86, 95
Cowpastures Bridge, 56, 75, 76, 86
Crown Hotel, 26, 44
Darling Harbour, 27
Elderslie, 18, 21, 28, 29, 33, 43, 44, 47, 48, 49, 57, 75, 76, 81, 84, 96
Elderslie High School, 81
Ellis Lane, 100
Exeter St., 36, 37, 83, 85, 92
Gladesville, 73
Glebe, 19
Great South Road, 95, 97
Grove Road, 21, 60, 85
Guangdong Province, 17, 19, 55, 66, 75, 92
Guangzhou, 17, 66, 79
Haymarket, 19, 44, 89, 92
Hilder Street, 21
Hop Chong garden, 42, 70, 73, 76, 80, 81
Hume Highway, 21, 49
John Street, ii, 44
Jung Seng, 19, 66
Kiangsi, 86
Little Sandy, 44
Macquarie Grove, 21, 40, 44, 46, 60, 61, 62, 85, 87, 89
Macquarie Grove Bridge, 21, 40, 61
Macarthur Bridge, 21, 34, 40, 43, 75, 84
Macarthur Road, 34
Milk Depot, 55, 58
Miss Davies' Walk, 92

Index

Mitchell Street, 21, 26
Mount Hunter, 47, 48
Museum of Chinese in Australia, 66
Narellan Hotel, 28
Nepean River, 15, 17, 21, 24, 34, 42, 43, 75, 76, 79, 87, 95, 96, 97, 98, 99, 100
Newscastle, 25
Parramatta, 56
Picton, 39, 41, 60
Rideouts sawmill, 26
Rookwood Cemetery, 29, 49, 51, 55, 57, 60, 62, 73, 74, 75, 81, 103
Rose Bowl, 44
San Yeck Garden, 43
Shanghai, 107
Sickles Bridge, 28
Smithfield, 59
South Coast, 18, 25
St. John's Church, 80
Stuckey bakery, 39
Sun Chon Key's garden, 70
Sydney Hospital, 70
Tamworth, Nemingha, 67
The Old Oaks Road, 28
University of Sydney, xi, 103, 108
Whitemans, 76
Wilkinson Street, 21
Wire Lane, 28
Zengcheng, 19, 66
Zhongshan, 60
Zengcheng, 19, 66

www.ingramcontent.com/pod-product-compliance
Lightning Source LLC
Chambersburg PA
CBRC092339290426
44109CB00008B/165